Applied Mathematical Sciences
Volume 89

Editors
F. John   J.E. Marsden   L. Sirovich

Advisors
M. Ghil   J.K. Hale   J. Keller
K. Kirchgässner   B. Matkowsky
J.T. Stuart   A. Weinstein

Applied Mathematical Sciences

(continued following index)

Robert E. O'Malley, Jr.

Singular Perturbation Methods for Ordinary Differential Equations

With 64 Illustrations

Springer-Verlag
New York Berlin Heidelberg London Paris
Tokyo Hong Kong Barcelona Budapest

Robert E. O'Malley, Jr.
Department of Applied Mathematics
University of Washington
Seattle, WA 98195
USA

Editors

F. John
Courant Institute of
    Mathematical Sciences
New York University
New York, NY 10012
USA

J.E. Marsden
Department of
    Mathematics
University of California
Berkeley, CA 94720
USA

L. Sirovich
Division of
    Applied Mathematics
Brown University
Providence, RI 02912
USA

Mathematical Subject Classifications: 34E15, 34E20, 34E05

Library of Congress Cataloging-in-Publication Data
O'Malley, Robert E.
    Singular perturbation methods for ordinary differential equations
/ Robert E. O'Malley, Jr.
        p.   cm. — (Applied mathematical sciences : v. 89)
    Includes bibliographical references and index.
    ISBN 0-387-97556-X
    1. Differential equations.   2. Perturbation (Mathematics)
I. Title.   II. Series: Applied mathematical sciences
(Springer-Verlag New York Inc.) ; v. 89.
QA1.A647 vol. 89
[QA372]
515'.352 — dc20   91-15273

Printed on acid-free paper.

Photocomposed copy prepared from author's TEX file.
Printed and bound by R.R. Donnelley & Sons, Harrisonburg, VA.
Printed in the United States of America.

9 8 7 6 5 4 3 2 1

ISBN 0-387-97556-X Springer-Verlag New York Berlin Heidelberg
ISBN 3-540-97556-X Springer-Verlag Berlin Heidelberg New York

Preface

This book results from various lectures given in recent years. Early drafts were used for several single semester courses on singular perturbation methods given at Rensselaer, and a more complete version was used for a one year course at the Technische Universität Wien. Some portions have been used for short lecture series at Universidad Central de Venezuela, West Virginia University, the University of Southern California, the University of California at Davis, East China Normal University, the University of Texas at Arlington, Universitá di Padova, and the University of New Hampshire, among other places. As a result, I've obtained lots of valuable feedback from students and listeners, for which I am grateful. This writing continues a pattern. Earlier lectures at Bell Laboratories, at the University of Edinburgh and New York University, and at the Australian National University led to my earlier works (1968, 1974, and 1978). All seem to have been useful for the study of singular perturbations, and I hope the same will be true of this monograph. I've personally learned much from reading and analyzing the works of others, so I would especially encourage readers to treat this book as an introduction to a diverse and exciting literature.

The topic coverage selected is personal and reflects my current opinions. An attempt has been made to encourage a consistent method of approaching problems, largely through correcting outer limits in regions of rapid change. Formal proofs of correctness are not emphasized. Instead, some nontrivial applications are included. In a first reading, one might be advised to skip to a later section from time to time (depending on one's background and interests). Little effort has been made to provide an exhaustive list of references. There are simply too many relevant works, so I've merely tried to be representative, emphasizing textbooks to some extent. Despite the special perspective, I hope the book will prove useful in teaching readers how to solve applied problems in a variety of contexts.

Many individuals deserve thanks for helping me understand many aspects of asymptotic analysis and its applications. My presentation here has been influenced by their papers and sometimes by their answers to specific questions. This has been especially true recently concerning historical information used to prepare the Appendix. Others helped generously through encouragement, hospitality, and support. In particular, much of the research reported was supported by the National Science Foundation, the U. S. Army Research Office, the Air Force Office of Scientific Research, and the Technical University of Vienna. Peggy Lashway and Jacques LaForgue worked especially hard to help me prepare the manuscript.

To all, many thanks.

Robert E. O'Malley, Jr.
Seattle, Winter 1990

Contents

Chapter 1

Examples Illustrating Regular and Singular Perturbation Concepts

A. The Harmonic Oscillator: Low-Frequency Situation

Consider a linear spring-mass system with forcing, but without damping, and with a small spring constant. This yields the differential equation

$$y'' + \epsilon y = f(x)$$

for the displacement $y(x)$ as a function of time x, with the small positive parameter ϵ being the ratio of the spring constant to the mass of the spring. This should be solved on the semi-infinite interval $x \geq 0$, with both the initial displacement $y(0)$ and the initial velocity $y'(0)$ prescribed. The traditional approach to solving such initial value problems [cf. Boyce and DiPrima (1986)] is to note that for ϵ small the homogeneous equation has the slowly varying solutions $\cos(\sqrt{\epsilon}x)$ and $\sin(\sqrt{\epsilon}x)$ and to look for a solution through variation of parameters. Specifically, one sets $y(x) = v_1(x)\cos(\sqrt{\epsilon}x) + v_2(x)\sin(\sqrt{\epsilon}x)$, where $v_1'\cos(\sqrt{\epsilon}x) + v_2'\sin(\sqrt{\epsilon}x) = 0$ and $-\sqrt{\epsilon}v_1'\sin(\sqrt{\epsilon}x) + \sqrt{\epsilon}v_2'\cos(\sqrt{\epsilon}x) = f(x)$. Since $y(0) = v_1(0)$ and $y'(0) = \sqrt{\epsilon}v_2(0)$, solving for v_1' and v_2' and integrating provides the unique solution

$$y(x, \epsilon) = y(0)\cos(\sqrt{\epsilon}x) + \frac{1}{\sqrt{\epsilon}}y'(0)\sin(\sqrt{\epsilon}x)$$

$$- \frac{1}{\sqrt{\epsilon}}\cos(\sqrt{\epsilon}x)\int_0^x \sin(\sqrt{\epsilon}t)f(t)dt$$

$$+ \frac{1}{\sqrt{\epsilon}}\sin(\sqrt{\epsilon}x)\int_0^x \cos(\sqrt{\epsilon}t)f(t)dt.$$

This result can be checked directly and confirmed to be the solution over any x interval. We note that it is sometimes convenient to represent the response for $y(0) = y'(0) = 0$ as the particular solution

$$y_p(x) = \int_0^x G(x, t, \epsilon)f(t)dt$$

and to call the kernel, $G(x, t, \epsilon) = (1/\sqrt{\epsilon})\sin[\sqrt{\epsilon}(x - t)]$, the Green's function for the initial value problem [cf., e.g., Roach (1982) or Stakgold (1979)].

Knowing this exact solution, unfortunately, does not conveniently display its behavior as $\epsilon \to 0$. We shall find the limiting behavior by using Taylor series expansions of the trig functions about $\epsilon = 0$, noting that their radii of convergence are infinite. Formally, then,

$$y(x, \epsilon) = \left(1 - \frac{\epsilon}{2}x^2 + \cdots\right)y(0) + \frac{1}{\sqrt{\epsilon}}\left(\sqrt{\epsilon}x - \frac{(\sqrt{\epsilon})^3}{6}x^3 + \cdots\right)y'(0)$$

$$+ \frac{1}{\sqrt{\epsilon}}\int_0^x \left(\sqrt{\epsilon}(x - t) - \frac{(\sqrt{\epsilon})^3}{6}(x - t)^3 + \cdots\right)f(t)dt,$$

so

$$y(x, \epsilon) = \left(y(0) + y'(0)x + \int_0^x (x - t)f(t)dt\right)$$

$$+ \epsilon\left(-\frac{y(0)x^2}{2} - \frac{y'(0)x^3}{6} - \int_0^x \frac{(x - t)^3}{6}f(t)dt\right) + \epsilon^2(\cdots).$$

Thus, the solution has a power series expansion in ϵ and a remainder after any number (say, $N + 1$) of terms could be obtained explicitly or could be estimated. In particular, for x on any *bounded* interval, that remainder is bounded above in absolute value by a term $B_N \epsilon^{N+1}$ with $0 < B_N < \infty$, for ϵ small enough. The series obtained is said to be an asymptotic expansion of the solution and to converge asymptotically to it [cf. Erdélyi (1956) or Olver (1974) for detailed presentations discussing the important, and somewhat mysterious, concept of an asymptotic approximation].

In brief, we shall say that a function $f(\epsilon)$ has the *asymptotic power series* expansion

$$f(\epsilon) \sim \sum_{j=0}^{\infty} f_j \epsilon^j \qquad \text{as } \epsilon \to 0$$

in the sense of Poincaré if, for any integer $N \geq 0$,

$$\frac{1}{\epsilon^N}\left(f(\epsilon) - \sum_{j=0}^{N} f_j \epsilon^j\right) \to 0 \qquad \text{as } \epsilon \to 0.$$

Often, we will have the somewhat stronger result that

$$\frac{1}{\epsilon^{N+1}}\left(f(\epsilon) - \sum_{j=0}^{N} f_j \epsilon^j\right)$$

will be bounded above and below for ϵ sufficiently small, and we will denote this by writing

$$f(\epsilon) = \sum_{j=0}^{N} f_j \epsilon^j + O(\epsilon^{N+1}) \qquad \text{as } \epsilon \to 0$$

calling O the "big O" Landau order symbol. We note that the coefficients f_j in such expansions are uniquely determined by the behavior of $f(\epsilon)$ as $\epsilon \to 0$. Even approximations that are limited to a finite N value can be of great practical use (even when the infinite series actually diverges!). Moreover, it is often helpful to generalize our concept to allow asymptotic sequences $\{\varphi_j(\epsilon)\}$ to replace the simple powers $\{\epsilon^j\}$ or to sometimes allow the coefficients f_j to also depend upon ϵ. We observe that, before Poincaré, asymptotic approximations were not always held in high regard [cf. Kline (1972)]. Abel is said to have written "Divergent series are the invention of the devil, and it is shameful to base on them any demonstration whatso-ever." We shall, nonetheless, proceed with confidence.

For ϵ small, we expect that the solution to our spring-mass problem will be well approximated by the solution of the limiting problem obtained when $\epsilon = 0$, i.e., by the function y_0 defined by

$$y_0'' = f(x), \quad y_0(0) = y(0), \quad y_0'(0) = y'(0)$$

(which has its own interpretation in terms of the deflection of a forced mass, without any spring). Its solution is obtained by integrating twice, yielding

$$y_0(x) = y(0) + y'(0)x + \int_0^x \int_0^t f(s)\,ds\,dt$$

or, reversing the order of integration,

$$y_0(x) = y(0) + y'(0)x + \int_0^x (x - s)f(s)\,ds.$$

This agrees with the leading term $y(x,0)$ of the asymptotic solution already obtained. More generally, if we instead look directly for a power series solution in the form

$$y(x, \epsilon) = y_0(x) + \epsilon y_1(x) + \epsilon^2 y_2(x) + \cdots,$$

the differential equation and the initial conditions imply that

$$(y_0'' + \epsilon y_1'' + \epsilon^2 y_2'' + \cdots) + \epsilon(y_0 + \epsilon y_1 + \epsilon^2 y_2 + \cdots) = f(x),$$

$$y_0(0) + \epsilon y_1(0) + \epsilon^2 y_2(0) + \cdots = y(0),$$

$$y_0'(0) + \epsilon y_1'(0) + \epsilon^2 y_2'(0) + \cdots = y'(0).$$

Successively equating coefficients of like powers ϵ^j of ϵ to zero yields

$$y_0'' = f(x), \quad y_0(0) = y(0), \quad y_0'(0) = y'(0)$$

and, for each $j > 0$,

$$y_j'' + y_{j-1} = 0, \quad y_j(0) = 0, \quad y_j'(0) = 0.$$

Thus, $y_0(x)$ is the unique solution of the limiting problem and, for each $j > 0$, we will have

$$y_j(x) = -\int_0^x (x-s)y_{j-1}(s)ds.$$

In particular, note that

$$y_1(x) = -\int_0^x (x-s)\left(y(0) + y'(0)s + \int_0^s (s-t)f(t)dt\right)ds$$

agrees with the term $y_1(x) = -y(0)\left(\frac{1}{2}x^2\right) - y'(0)\left(\frac{1}{6}x^3\right) - \int_0^x (x-t)^3\left[\frac{1}{6}f(t)\right]dt$ found previously. Indeed, the definition of a Taylor series for y about $\epsilon = 0$ implies that we must uniquely obtain

$$y_j(x) = \frac{1}{j!}\frac{d^j}{d\epsilon^j}y(x,\epsilon)\Big|_{\epsilon=0} \qquad \text{for each } j \geq 0,$$

presuming the solution is infinitely differentiable in ϵ. We point out that this formula holds for asymptotic, as well as convergent, series, though the derivatives naturally have to be replaced by appropriate one-sided limits when $\epsilon \geq 0$. Our procedure is, indeed, also correct for obtaining solutions for small negative values of ϵ, though that expansion will not have any physical interpretation concerning a spring and mass, nor will the original derivation in terms of $\cos(\sqrt{\epsilon}x)$ and $\sin(\sqrt{\epsilon}x)$ then be appropriate. Finally, we note that the expansion must break down as $x \to \infty$, because the approximation $y_0(x)$ (like later finite-term approximations) fails to display the long-term oscillatory behavior [just like 1 or $1 - \epsilon x^2/2$ become poor approximations to $\cos(\sqrt{\epsilon}x)$; this is made eminently clear if we note that the first correction term $-\epsilon x^2/2$ to the limit 1 dominates it when $x > \sqrt{2/\epsilon}$]. We might, of course, anticipate complications in our model as we let $x \to \infty$, since we have not introduced any damping, which is necessary to provide the physical system asymptotic stability.

B. Introductory Definitions and Remarks

We give the following informal definition:

A *regular perturbation* problem $P_\epsilon(y_\epsilon) = 0$ depends on its small parameter ϵ in such a way that its solution $y_\epsilon(x)$ converges as $\epsilon \to 0$ (uniformly with respect to the independent variable x in the relevant domain) to the solution $y_0(x)$ of the limiting problem $P_0(y_0) = 0$. An example is given by the spring-mass problem on any finite-time interval.

The parameter ϵ typically represents the influence of many nearly negligible physical influences. Usually, we will restrict attention to boundary value problems where P_ϵ is defined by differential operators and boundary forms, though one might also study integral or other operator equations or more global auxiliary conditions. Assuming sufficient smoothness (with respect to y, x, and ϵ), the solution of a regular perturbation problem can be approximated by a formal asymptotic power series expansion in ϵ having the leading term (i.e., asymptotic limit) y_0.

A *singular perturbation* is said to occur whenever the regular perturbation limit $y_\epsilon(x) \to y_0(x)$ fails. Such a breakdown, typically, only occurs in narrow intervals of space or short intervals of time (although secular problems with nonuniform behavior at infinity, as we found for the harmonic oscillator, are also common). Much of the special vocabulary of singular perturbations comes from physically natural terminology in high Reynolds number fluid flow past physical bodies [cf., e.g., Van Dyke (1964) and the Appendix]. In such a flow, a *no-slip* condition along the surface results in a thin *boundary layer* of nonuniform convergence about the body where the velocity varies from zero to that of the uniform outside flow. The parameter ϵ is typically the reciprocal of the nondimensional Reynolds number Re, defined to be UL/ν with U being the velocity of the uniform flow; L, the characteristic length of the body; and ν, the viscosity coefficient per unit fluid density. This application was first systematically analyzed by Prandtl (1905), but the common occurrence of physically meaningful (usually dimensionless) small or large parameters in many scientific studies often calls for a corresponding singular perturbation analysis [cf. Lin and Segel (1974) for an elementary discussion of scaling, dimensional analysis, and singular perturbations].

C. A Simple First-Order Linear Initial Value Problem

Consider the problem

$$\epsilon \dot{x} + x = 1, \quad \text{with } x(0) \text{ prescribed, for } t \geq 0$$

where ϵ is a small *positive* parameter. We might anticipate difficulties in solving the problem on any fixed t interval because the relevant Lipschitz constant $1/\epsilon$, which is assumed to be bounded in most existence–uniqueness theories [cf. Coddington and Levinson (1955)] and even in most analyses of finite-difference methods [cf. Henrici (1962)], now becomes unbounded as $\epsilon \to 0$. We note that the presence of a small parameter before the highest derivative(s) in a differential equation, though not necessary to obtain a singular perturbation problem, commonly signals the possibility of such. Our equation can be immediately integrated, using the integrating factor $e^{t/\epsilon}$. This yields the solution

$$x(t, \epsilon) = 1 + [x(0) - 1]e^{-t/\epsilon}.$$

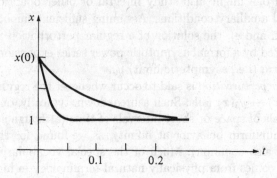

Fig. 1.1. $x(t, \epsilon) = 1 + 2e^{-t/\epsilon}$ for $\epsilon = 0.1$ and 0.01.

For $\epsilon < 0$ and $x(0) \neq 1$, the solution will become unbounded as $\epsilon \to 0$ for any $t > 0$. For $\epsilon > 0$, however, the solution tends to 1 for any $t > 0$ as $\epsilon \to 0$. Graphing the solution $x(t, \epsilon)$, for example, for $x(0) = 3$ and small ϵ, we obtain Figure 1.1.

As $\epsilon \to 0$, $x(t, \epsilon)$ will increase monotonically toward the constant limit 1, for each $t > 0$, if $x(0) < 1$. Provided $x(0) \neq 1$, then $x(t, \epsilon)$ has a discontinuous limit as $\epsilon \to 0$, namely,

$$x(t, \epsilon) \to \begin{cases} x(0) & \text{for } t = 0 \\ 1 & \text{for } t > 0. \end{cases}$$

This shows that convergence is not uniform at $t = 0$. (Recall that the uniform limit of continuous functions is continuous.) The region of nonuniform convergence is said to be $O(\epsilon)$ thick, since for $t > B\epsilon$, $|x(t, \epsilon) - 1| \leq |x(0) - 1|e^{-B}$ can be made arbitrarily close to zero by picking $B > 0$ large enough. We will call the interval of nonuniform convergence an *initial layer*.

Because of the observed need to distinguish between $\epsilon \to 0^-$ and $\epsilon \to 0^+$, we shall *henceforth* always let ϵ represent a *small positive parameter*. It is important to observe that the exact solution to this singular perturbation problem is the sum of (i) a function of the independent variable t (actually, a constant function) and (ii) a function of the "stretched time"

$$\tau = \frac{t}{\epsilon}$$

which decays to zero as $\tau \to \infty$. The former "outer solution"

$$X(t, \epsilon) = 1$$

is a smooth solution of the differential equation which provides the asymptotic solution [to arbitrary orders $O(\epsilon^N)$] for $t > 0$, and the latter "initial layer correction"

$$\xi(\tau, \epsilon) = e^{-\tau}[x(0) - 1]$$

provides the needed nonuniform convergence in the narrow "boundary layer" region near $t = 0$. The reason for introducing the stretching τ is to provide ourselves a microscope which magnifies the $O(\epsilon)$ thick initial layer into the semi-infinite interval $\tau \geq 0$, thereby disclosing the nonuniform convergence. We note that much more general "two-timing" methods to obtain asymptotic solutions to nonlinear problems are presented in Kevorkian and Cole (1981) and elsewhere.

An alternative solution technique is to separately solve the problem in the "inner" region $\tau \geq 0$ and in the "outer" region $t > 0$ and to "match" them at the edge of the boundary layer. [Friedrichs (1952) also discussed "patching" solutions together, as might be attempted at a small t value like $-10\epsilon \log \epsilon$.] Let us first seek an inner solution $\tilde{x}(\tau, \epsilon)$ on $\tau \geq 0$. The inner problem takes the form

$$\frac{d\tilde{x}}{d\tau} = -(\tilde{x} - 1), \qquad \tilde{x}(0) = x(0).$$

For all ϵ values, it has the solution

$$\tilde{x}(\tau, \epsilon) = 1 + (x(0) - 1)e^{-\tau}$$

on $\tau \geq 0$. In particular, note that its steady state is $\tilde{x}(\infty, \epsilon) = 1$. Beyond the initial layer (i.e., for $t > 0^+$), let us seek an outer solution $X(t, \epsilon)$ as a smooth solution of

$$\epsilon \dot{X} + X = 1$$

with an asymptotic power series expansion

$$X(t, \epsilon) \sim \sum_{j=0}^{\infty} X_j(t)\epsilon^j.$$

Formally equating coefficients of like powers of ϵ on both sides of the differential equation, we successively require that

$$X_0 = 1$$

and, for each $j > 0$,

$$X_j = -\dot{X}_{j-1}.$$

Thus, $X_j(t) = 0$ for each $j > 0$, so the outer solution

$$X(t, \epsilon) = X_0(t) = 1$$

is constant in both t and ϵ. The essential point is that the inner solution $\tilde{x}(\tau, \epsilon)$ and the outer solution $X(t, \epsilon)$ actually match at the edge of the initial layer since $\lim_{\tau \to \infty} \tilde{x}(\tau, \epsilon) = 1 = \lim_{t \to 0} X(t, \epsilon)$. We observe that more complicated matching is sometimes necessary; see, for example, Kevorkian and Cole's (1981) discussion of limit process expansions [for more basic explanations and examples, consult Fraenkel (1969), Eckhaus (1979), Lagerstrom (1988), and Il'in (1989)].

We note that the outer limit, $X_0(t) = 1$ for $t > 0$, solves the reduced problem $X_0 = 1$. Unless $x(0) = 1$, the discontinuous limit implies that convergence at $t = 0$ will be nonuniform. One motivation for introducing $\tau = t/\epsilon$ as the stretched time is that the change of time scales from t to τ balances the bounded size of the terms in the differential equation. Another is that the characteristic polynomial $\epsilon\lambda + 1 = 0$ for the homogeneous differential operator has the large (stable) root $\lambda = -1/\epsilon$ (in the left half-plane), so the complementary solutions $x_0(t) = e^{-t/\epsilon}c_0$ evolve on an $O(\epsilon)$ t-scale. It is important to realize that such initial value problems are not straightforward to integrate numerically. (We can only partly cure our asymptotic problems by using a bigger computer.) If, for example, one tried to integrate $\epsilon\dot{x} + x = 1$ on $0 \leq t \leq 1$ using Euler's method with N equal time steps h, the resulting difference equation

$$\epsilon\left(\frac{x_{n+1} - x_n}{h}\right) + x_n = 1$$

would have the solution

$$x_{n+1} = 1 + (x_0 - 1)\left(1 - \frac{h}{\epsilon}\right)^{n+1}.$$

Convergence for large n requires us to take $|1 - h/\epsilon| \leq 1$. Thus, for ϵ small, we need $h < 2\epsilon$. This is very restrictive because roughly $1/\epsilon$ time steps will be needed to reach $t = 1$, even though the exact solution is nearly constant outside the $O(\epsilon)$ thin initial layer of nonuniform convergence. This situation is typical of *stiff* differential equations, where the step size is restricted by stability, rather than accuracy, considerations [cf. Aiken (1985)]. The problem can be avoided by using so-called "upwinding," i.e., a backward difference formula for the derivative, or by using an appropriate nonuniform mesh which concentrates meshpoints in the initial layer of rapid change. (How to do so may take some skill [cf. Pearson (1968)].)

Exercises

1. [Compare Simmonds and Mann (1986)]
    Determine asymptotic approximations to the roots of the following polynomials (as the parameters ϵ and μ tend to zero):

    (a) $x^2 + x + \epsilon = 0$,   (e) $x^2 + \epsilon x + \mu = 0$,

    (b) $\epsilon x^2 + x + 1 = 0$,   (f) $\epsilon x^2 + x + \mu = 0$,

    (c) $x^3 + x^2 - \epsilon = 0$,   (g) $\epsilon x^2 + \mu x + 1 = 0$,

    (d) $\epsilon x^4 - x^2 + 1 = 0$.

2. [Compare O'Malley (1988)]

Consider the initial value problem

$$\epsilon \dot{y} = y - y^3, \quad y(0) \text{ prescribed.}$$

a. Find a formula for the solution by rewriting the equation as a Riccati differential equation for y^2. (The equation is also separable.)

b. Determine the range of initial values $y(0)$ which will lead to the limiting solution $Y_0(t) = -1$ on $t > 0$. (Hint: Find the steady states of the inner problem.) An alternative solution method would be to note that $\epsilon \dot{y} = (1-y)y(y+1)$, so the sign of $\dot{y}$ on various y intervals completely specifies the limiting behavior without the intermediate step of obtaining an explicit solution.

3.

a. Consider the two-point problem

$$\begin{cases} \epsilon \ddot{x} + \dot{x} + x = 0, & 0 \le t \le 1, \\ x(0) = 0, & x(1) = 1. \end{cases}$$

Determine the exact solution and show that

$$x(t, \epsilon) \sim e^{1-t} - e^{1-t/\epsilon} \quad \text{as } \epsilon \to 0$$

uniformly in $0 \le t \le 1$. [This example is often called Friedrichs' problem. Friedrichs' contributions to singular perturbations (cf. Appendix) were, indeed, much more substantial than the example alone suggests.]

b. Find the initial value $x(0, \epsilon)$ in order that the two-point problem

$$\begin{cases} \epsilon \ddot{x} + \dot{x} + x = 0 \\ x(0, \epsilon) \text{ as determined}, \quad x(1) = 1, \end{cases}$$

has a smooth solution with an asymptotic power series expansion $x(t, \epsilon) \sim \sum_{j=0}^{\infty} x_j(t) \epsilon^j$ uniformly valid throughout $0 \le t \le 1$. Note that $x(t, \epsilon)$ will then agree asymptotically with its outer expansion, so the initial layer correction can be eliminated by the "right" selection, $x(0, \epsilon) = \exp[(1 - \sqrt{1 - 4\epsilon})/2\epsilon]$.

4. [Compare O'Malley (1974)]

a. Consider the linear initial value problem

$$\begin{cases} \epsilon \ddot{x} + \mu \dot{x} + x = 0, & t \ge 0, \\ x(0), & \dot{x}(0) \text{ prescribed} \end{cases}$$

when $\epsilon/\mu^2 \to 0$ as $\mu \to 0$. Show why it is natural to seek a solution in the form

$$x(t, \epsilon, \mu) = e^{-q_1 t/\mu} k_1 + \frac{\epsilon}{\mu^2} e^{-q_2 t\mu/\epsilon} k_2$$

where $q_1(\epsilon/\mu^2)$ and $q_2(\epsilon/\mu^2)$ have power series expansions in ϵ/μ^2 and $k_1(\epsilon/\mu^2, \mu)$ and $k_2(\epsilon/\mu^2, \mu)$ have double power series expansions in the two small parameters ϵ/μ^2 and μ.

b. Show that (as $\epsilon/\mu^2 \to 0$)

$$x(t, \epsilon, \mu) \sim e^{-t/\mu}x(0)$$

and that

$$\mu\dot{x}(t, \epsilon, \mu) \sim (e^{-t/\mu} - e^{-t\mu/\epsilon})x(0),$$

so

$$\dot{x}(t, \epsilon, \mu) = O\left(\frac{1}{\mu}e^{-t/\mu}\right).$$

(As $\mu \to 0$, then $\dot{x}$ behaves like a delta function peaked at $t = 0$.)

c. Explain why a limiting solution will not generally exist if, instead, $\mu^2/\epsilon \to 0$ as $\epsilon \to 0$. [We note that two-parameter problems are considered extensively in O'Malley (1974), and that a quite different approach to multiparameter problems occurs in Khalil (1981) and Khalil and Kokotovic (1979).]

D. Some Second-Order Two-Point Problems

(i) If we consider the free motion of the undamped linear spring-mass system with a very resistant spring, we can expect rapidly varying solutions (indeed, high-frequency oscillations) to result. Let us prescribe specific displacements at times $t = 0$ and 1 to obtain the two-point problem

$$\epsilon^2\ddot{x} + x = 0, \quad 0 \leq t \leq 1, \quad x(0) = 0, \quad x(1) = 1,$$

where ϵ^2, the ratio of the mass to the spring constant, is small. Corresponding to the large imaginary roots $\pm i/\epsilon$ of the relevant characteristic polynomial are the linearly independent rapidly oscillating solutions $\sin(t/\epsilon)$ and $\cos(t/\epsilon)$ of the differential equation. The boundary value problem, then, has the unique solution

$$x(t, \epsilon) = \frac{\sin(t/\epsilon)}{\sin(1/\epsilon)}$$

for $\epsilon \neq 1/n\pi$, $n = 1, 2, \ldots$. For nonexceptional small positive values of ϵ, the ratio oscillates rapidly, so no pointwise limit exists as $\epsilon \to 0$ (the average is zero, however). For the exceptional values of ϵ, the problem has no solutions.

(ii) The asymptotic behavior of solutions to the superficially similar problem

$$\epsilon^2\ddot{x} - x = 0, \quad x(0) = 0, \quad x(1) = 1$$

on $0 \leq t \leq 1$ is much less complicated. Here, $e^{\pm t/\epsilon}$ are linearly independent solutions of the equation (one decaying exponentially and the other growing exponentially within $0 < t < 1$). With some foresight, we might recognize the comparable roles of the endpoints $t = 0$ and 1, and might be motivated to seek solutions in the special form

$$x(t, \epsilon) = e^{-t/\epsilon}a + e^{-(1-t)/\epsilon}b,$$

where a and b are independent of t and are yet undetermined as functions of ϵ. Since the values of the functions $e^{-t/\epsilon}$ and $e^{-(1-t)/\epsilon}$ both lie between 0 and 1, the first describes typical initial layer behavior with nonuniform convergence to 0 in a right-sided $O(\epsilon)$ neighborhood of $t = 0$ and the latter does likewise in a left-sided $O(\epsilon)$ neighborhood of $t = 1$. The boundary conditions require that $0 = a + e^{-1/\epsilon}b$ and $1 = e^{-1/\epsilon}a + b$, so

$$a = -\frac{e^{-1/\epsilon}}{1 - e^{-2/\epsilon}}$$

and

$$b = \frac{1}{1 - e^{-2/\epsilon}}$$

are uniquely determined. For ϵ small, $e^{-1/\epsilon}$ is asymptotically negligible (i.e., smaller than any power ϵ^k with $k > 0$), so $a \sim 0$ and $b \sim 1$ and we have shown that a uniformly valid asymptotic solution is given by

$$x(t, \epsilon) \sim e^{-(1-t)/\epsilon}$$

with an error smaller than $O(\epsilon^{N+1})$ for any $N > 0$.

We note that one should not always throw away asymptotically negligible terms. The use of "asymptotics beyond all orders" is critical in many exciting new applications. For arbitrary boundary values $x(0)$ and $x(1)$, the asymptotic solution is uniquely given by

$$x(t, \epsilon) \sim e^{-t/\epsilon}x(0) + e^{-(1-t)/\epsilon}x(1).$$

For a small positive value of ϵ, a typical solution is presented in Figure 1.2. Within $(0, 1)$, the limiting ("outer") solution is the smooth trivial solution $X(t, \epsilon) = 0$ of the differential equation [correct to all orders $O(\epsilon^N)$ as a formal power series expansion procedure would readily show]. The uniformly valid solution is, indeed, best described in terms of the stretched variables $\kappa = t/\epsilon$ and $\lambda = (1 - t)/\epsilon$ and written as

$$x(t, \epsilon) = e^{-\kappa}x(0) + e^{-\lambda}x(1).$$

(For more general problems, we would also have t dependence.) For any fixed $t < 1$, $\lambda \sim \infty$, so $e^{-\lambda}$ is asymptotically negligible and

$$x(t, \epsilon) \sim e^{-\kappa}x(0),$$

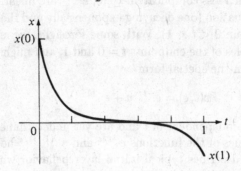

Fig. 1.2. $x(t,\epsilon) \sim e^{-t/\epsilon}x(0) + e^{-(1-t)/\epsilon}x(1)$ for $x(0) = 5$, $x(1) = -2$, and $\epsilon = 0.1$.

whereas

$$x(t,\epsilon) \sim e^{-\lambda}x(1) \qquad \text{for } 0 < t \leq 1$$

and

$$x(t,\epsilon) \sim 0 \qquad \text{within } 0 < t < 1.$$

Note that this trivial limiting solution satisfies the reduced problem obtained by setting $\epsilon = 0$.

(iii) For a nonlinear example, consider the Dirichlet problem for

$$\epsilon\ddot{x} + x\dot{x} = 0 \quad \text{on } 0 \leq t \leq 1$$

[cf. Wasow (1970) and O'Malley (1974)] where $x(0)$ and $x(1)$ are prescribed. It could describe the motion of a mass moving in a medium with damping proportional to the displacement, with either the mass small or the damping large. Here, the limiting equation $X_0\dot{X}_0 = 0$ suggests that all limiting solutions will be constants. One might further guess that the most likely constant limits would be the values $x(0), x(1)$, and 0. {Differential inequality and maximum principle arguments [cf. Chang and Howes (1984), Dorr et al. (1973), and Section 3G] can actually be used to show that the problem has a unique solution (for every ϵ) which is bounded between $x(0)$ and $x(1)$.} Kreiss and Lorenz (1989) also analyze this steady-state Burgers' equation and various generalizations.

We will construct asymptotic solutions, showing that their structure differs considerably in four cases, depending on the particular end values $x(0)$ and $x(1)$.

Case 1: If $x(1) > 0$ and $x(0) > -x(1)$, let us look for a solution in the form

$$x(t,\epsilon) = x(1) + \eta(\tau),$$

where $\eta \to 0$ as $\tau = t/\epsilon \to \infty$. (Motivation for this choice might be provided by the simple linear equation $\epsilon\ddot{x} + \alpha\dot{x} = 0$ which has an asymptotic

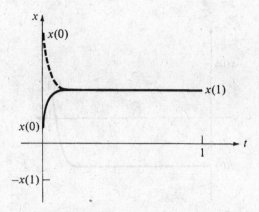

Fig. 1.3. Solution with an initial layer.

solution with initial layer behavior provided $\alpha > 0$.) Clearly, this ansatz (or presumed form of solution) implies that the solutions for ϵ small should be as shown in Figure 1.3. There is a limiting outer solution $x(1)$ and an initial layer of nonuniform convergence of $O(\epsilon)$ thickness in t, described by the function $\eta(\tau)$. The representation requires η to satisfy the stretched problem

$$\frac{d^2\eta}{d\tau^2} + (x(1) + \eta)\frac{d\eta}{d\tau} = 0$$

on the semi-infinite interval $\tau \geq 0$ with the boundary conditions $\eta(0) = x(0) - x(1)$ and $\eta(\infty) = 0$. Integrating backward from infinity shows that η must satisfy an initial value problem on $\tau \geq 0$ for the Riccati equation

$$\frac{d\eta}{d\tau} + x(1)\eta + \frac{1}{2}\eta^2 = 0.$$

This autonomous equation has rest points 0 and $-2x(1)$ which are, respectively, stable and unstable because $x(1) > 0$. (Simply check the sign of $d\eta/d\tau$ for all values of η.) Thus, η will decay monotonically to zero if $\eta(0) > 0$ and it will increase monotonically to zero if $-2x(1) < \eta(0) < 0$, i.e., if $-x(1) < x(0) < x(1)$. If, however, $\eta(0) < -2x(1)$, i.e., $x(0) < -x(1)$, $\eta(\tau)$ will become unbounded for some finite τ (so our ansatz is not appropriate). Closed-form expressions for these solutions could be given, but they are unnecessary. Note that this analysis provides the existence and asymptotic behavior of the unique solution to our boundary value problem provided the boundary values lie in the first $135°$ sector of the $x(0) - x(1)$ plane.

Case 2: In completely analogous fashion, we could show that the unique solution takes the form

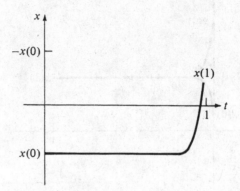

Fig. 1.4. Solution with a terminal layer.

$$x(t, \epsilon) = x(0) + \xi(\sigma),$$

where $\xi \to 0$ as the stretched variable $\sigma = (1-t)/\epsilon \to \infty$, provided $x(0) < 0$ and $x(1) < -x(0)$. Pictorially, the solution $x(t, \epsilon)$ for ϵ small will appear as in Figure 1.4. Thus, there is now the constant limiting solution $x(0)$ except in a terminal layer of $O(\epsilon)$ thickness. Thus far, then, the limiting solution has been determined in three-fourths of the $x(0) - x(1)$ plane.

Case 3: On the borderline between Cases 1 and 2, we will have $x(0) = -x(1) < 0$. Symmetry suggests that we might then expect $x(\frac{1}{2}, \epsilon) = 0$. Let us seek a solution

$$x(t, \epsilon) = \rho(\kappa)$$

for $-\infty < \kappa = (t - \frac{1}{2})/\epsilon < \infty$ such that $\rho(0) = 0$, $\rho \to x(0)$ as $\kappa \to -\infty$ and $\rho \to x(1)$ as $\kappa \to \infty$. We'll then have nonuniform convergence in an $O(\epsilon)$ thick layer about $t = \frac{1}{2}$. Pictorially, for small ϵ, we will obtain Figure 1.5. Such an interior shock or transition layer is typical of important phenomena in gas dynamics and other applied fields [cf. Smoller (1983) and Fife (1988)]. Looking backward and forward from $t = \frac{1}{2}$, the shock layer appears to be a concatenation of a terminal layer and an initial layer, which were both previously studied.

The differential equation for x implies that we will need

$$\frac{d^2 \rho}{d\kappa^2} + \rho \frac{d\rho}{d\kappa} = 0$$

on $-\infty < \kappa < \infty$. (This stretched equation is now no simpler than the original one.) Integrating from either $\pm\infty$ then implies that

$$\frac{d\rho}{d\kappa} + \frac{1}{2}\rho^2 = \frac{1}{2}x^2(0).$$

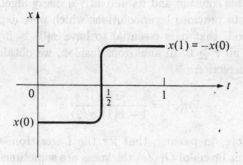

Fig. 1.5. Solution with a shock layer.

Because $\rho(0) = 0$, we will obtain the unique monotonically increasing solution

$$\rho(\kappa) = x(0) \tanh\left[\frac{1}{2}x(0)\kappa\right].$$

Because $x(0) < 0$, this solution displays the desired behavior as κ tends to both $\pm\infty$. We note that for the related, but more complicated, Lagerstrom–Cole equation $\epsilon\ddot{x} + x\dot{x} - x = 0$ (which cannot be integrated explicitly) the location of the shock layer will instead vary with the boundary values [cf. Kevorkian and Cole (1981) and Lagerstrom (1988)].

Case 4:   Finally, in the quarter-plane where $x(1) < 0 < x(0)$, we look for a solution of the form

$$x(t, \epsilon) = \xi(\tau, \epsilon) + \eta(\sigma, \epsilon),$$

where $\xi \to 0$ as $\tau = t/\epsilon \to \infty$ and $\eta \to 0$ as $\sigma = (1 - t)/\epsilon \to \infty$. If we are successful, then the asymptotically valid outer solution within $0 < t < 1$ will be trivial (to all orders in ϵ) and there will be $O(\epsilon)$ thick boundary layers of nonuniform convergence near both endpoints. In particular, near $t = 0$, η and its derivatives will be asymptotically negligible, so $d^j x(t, \epsilon)/dt^j \sim (1/\epsilon^j)[d^j\xi(\tau, \epsilon)/d\tau^j]$ locally, and the initial boundary layer correction $\xi(\tau, \epsilon)$ will necessarily satisfy

$$\frac{d^2\xi}{d\tau^2} + \xi\frac{d\xi}{d\tau} = 0.$$

Integrating backward from infinity,

$$\frac{d\xi}{d\tau} + \frac{1}{2}\xi^2 = 0.$$

Since $\xi(0) \sim x(0)$, we finally obtain

$$\xi(\tau) = \frac{x(0)}{1 + x(0)(\tau/2)}.$$

Note, first, that this solution and its derivatives decay algebraically at infinity, in contrast to previous layer solutions which were exponentially decaying, and, second, that it is essential to have $x(0) > 0$ in order for ξ to be defined for all $\tau \geq 0$. In analogous fashion, we obtain the terminal boundary layer correction

$$\eta(\sigma) = \frac{x(1)}{1 - x(1)(\sigma/2)}$$

on $\sigma \geq 0$. We note, in passing, that for the Lagerstrom–Cole example, additional endpoint layers of $O(\sqrt{\epsilon})$ thickness are sometimes needed as intermediates between the $O(\epsilon)$ thick boundary layers and the outer solution.

For our example, the limiting interior solution is always one of the three predicted constants $x(0), x(1)$, or 0. Indeed, we have shown that

$$x(t) \sim \begin{cases} x(1) \text{ for } t > 0 & \text{if } 0 \leq \arg[x(1)/x(0)] < 3\pi/4 \\ \left\{ \begin{array}{l} x(0) \text{ for } t < \frac{1}{2} \\ x(1) \text{ for } t > \frac{1}{2} \end{array} \right\} & \text{if } \arg[x(1)/x(0)] = 3\pi/4 \\ x(0) \text{ for } t < 1 & \text{if } 3\pi/4 < \arg[x(1)/x(0)] \leq 3\pi/2 \\ 0 \text{ for } 0 < t < 1 & \text{if } 3\pi/2 \leq \arg[x(1)/x(0)] \leq 2\pi \,. \end{cases}$$

The limiting solutions, then, vary smoothly in the $x(0) - x(1)$ plane, except near the 135° ray, where a small change in $x(0)$ or $x(1)$ will move the layer of nonuniform convergence from $t = \frac{1}{2}$ to either $t = 0$ or 1. Readers might want to experiment numerically with the solution as the end values vary with ϵ near the critical ray.

Exercise

[Compare Dieudonné (1973)]

Suppose X_0 is an isolated root of the nonlinear vector equation

$$f(X_0, 0) = 0$$

along which the Jacobian matrix $f_x(X_0, 0)$ is nonsingular. Assuming sufficient smoothness, show how to formally obtain a power series solution

$$X(\epsilon) = X_0 + \epsilon X_1 + \epsilon^2 X_2 + \cdots$$

of the perturbed equation

$$f(X, \epsilon) = 0$$

for ϵ small. Note that this result provides a constructive version of the implicit function theorem.

(iv) To illustrate the inherent difficulty of treating nonlinear (indeed, superquadratic) singular perturbation problems, Coddington and Levinson (1952) introduced the example

$$\begin{cases} \epsilon \ddot{x} + \dot{x} + \dot{x}^3 = 0, \\ x(0), x(1) \text{ being prescribed constants.} \end{cases}$$

If $x(0) = x(1)$, we have the constant solution. Otherwise, $y = \dot{x}$ will satisfy $\epsilon \dot{y} = -(y + y^3)$, so separation of variables implies that $y = \pm e^{-(t+\alpha)/\epsilon}/\sqrt{1 - e^{-2(t+\alpha)/\epsilon}}$ for some constant α, and a further integration yields $x(t) = x(0) + \int_0^t y(s)ds = x(0) \mp \epsilon \sin^{-1}(e^{-(t+\alpha)/\epsilon})$. For y to be bounded throughout $0 \leq t \leq 1$, we will need $t + \alpha > 0$ there. Then, however, $|y|$ will decay exponentially and x can only change by $O(\epsilon)$. When ϵ is small enough, this contradicts the fact that $x(0) \neq x(1)$. Alternatively, we can use the remaining boundary condition to eliminate α and obtain

$$e^{(t-1)/\epsilon} \sin\left(\frac{x(t) - x(0)}{\epsilon} \right) = \sin\left(\frac{x(1) - x(0)}{\epsilon} \right).$$

Consider $\epsilon_n = [2|x(1) - x(0)|]/\pi(1 + 4n)$, $n = 0, 1, 2, \ldots$. Then, we would need $e^{(t-1)/\epsilon_n} \sin\{[x(t) - x(0)]/\epsilon_n\} = 1$, $0 \leq t \leq 1$. This is impossible, so the boundary value problem cannot have a solution for all sufficiently small values of ϵ.

E. Regular Perturbation Theory for Initial Value Problems

Consider the vector initial value problem

$$\dot{x} = f(x, t, \epsilon), \quad x(0) = x_0(\epsilon)$$

where ϵ is a small parameter and suppose that the limiting problem

$$\dot{X}_0 = f(X_0, t, 0), \quad X_0(0) = x_0(0)$$

has a unique smooth (continuously differentiable) solution $X_0(t)$ throughout the interval $0 \leq t \leq T$. If $f(x, t, \epsilon)$ and $x_0(\epsilon)$ depend smoothly on their arguments, we should expect the solution $x(t, \epsilon)$ to depend smoothly on t and ϵ and to tend to the limit $X_0(t)$ on $0 \leq t \leq T$ as $\epsilon \to 0$. (We do not even need to restrict attention to positive values of ϵ.) Indeed, the same existence–uniqueness theory [cf. Coddington and Levinson (1955)] which guarantees the local existence of $X_0(t)$ (based on having a bounded Lipschitz constant) also shows continuous dependence on parameters which are here generically represented by the scalar ϵ. Formally, let us suppose that the asymptotic power series expansions

$$f(x, t, \epsilon) \sim \sum_{j=0}^{\infty} f_j(x, t)\epsilon^j$$

and

$$x_0(\epsilon) \sim \sum_{j=0}^{\infty} x_{0j}\epsilon^j$$

hold and let us look for an asymptotic power series solution

$$X(t, \epsilon) \sim \sum_{j=0}^{\infty} X_j(t)\epsilon^j$$

of the initial value problem. To do so, we need to obtain the Taylor series expansion for $f(X(t, \epsilon), t, \epsilon)$ about $x = X_0(t), t$, and $\epsilon = 0$ [cf., e.g., Apostol (1957)]. Up to second-order terms in $X - X_0$ and ϵ, we will have

$$f(X(t, \epsilon), t, \epsilon) = f(X_0(t), t, 0) + \left(\frac{\partial f}{\partial x}(X_0, t, 0)(X - X_0) + \epsilon \frac{\partial f}{\partial \epsilon}(X_0, t, 0) \right)$$

$$+ \frac{1}{2} \left[\left(\frac{\partial^2 f}{\partial x^2}(X_0, t, 0)(X - X_0) \right)(X - X_0) \right.$$

$$\left. + 2\epsilon \frac{\partial^2 f}{\partial x \partial \epsilon}(X_0, t, 0)(X - X_0) + \epsilon^2 \frac{\partial^2 f}{\partial \epsilon^2}(X_0, t, 0) \right] + \cdots$$

with the quadratic terms having their natural interpretations. Substituting the expansion for $X - X_0$ finally provides us a power series in ϵ for f. Thus, we have

$$f(X(t, \epsilon), t, \epsilon) = f_0(X_0, t) + \left(\frac{\partial f_0}{\partial x}(X_0, t)(\epsilon X_1 + \epsilon^2 X_2 + \cdots) + \epsilon f_1(X_0, t) \right)$$

$$+ \frac{1}{2} \left[\left(\frac{\partial^2 f_0}{\partial x^2}(X_0, t)(\epsilon X_1 + \epsilon^2 X_2 + \cdots) + 2\epsilon \frac{\partial f_1}{\partial x}(X_0, t) \right) \right.$$

$$\left. \times (\epsilon X_1 + \epsilon^2 X_2 + \cdots) + 2\epsilon^2 f_2(X_0, t) \right] + \cdots$$

$$= f_0(X_0, t) + \epsilon \left(\frac{\partial f_0}{\partial x}(X_0, t)X_1 + f_1(X_0, t) \right)$$

$$+ \epsilon^2 \left[\frac{\partial f_0}{\partial x}(X_0, t)X_2 + \frac{1}{2} \left(\frac{\partial^2 f_0}{\partial x^2}(X_0, t)X_1 \right) X_1 \right.$$

$$\left. + \frac{\partial f_1}{\partial x}(X_0, t)X_1 + f_2(X_0, t) \right] + O(\epsilon^3).$$

The jth coefficient in the resulting series must, of course, coincide with the Taylor series coefficient $(1/j!)(d^j/d\epsilon^j)f(X(t, \epsilon), t, \epsilon)|_{\epsilon=0}$. We observe that much of the labor involved in such expansion procedures can be efficiently handled by modern symbol manipulation routines like MACSYMA, REDUCE, or MAPLE [cf. Rand and Armbruster (1987) and Keener (1988)]. Equating like powers of ϵ, then, in both the differential system and the initial condition implies that we must successively have

$$\dot{X}_0 = f_0(X_0, t), \quad X_0(0) = x_{00},$$

$$\dot{X}_1 = \frac{\partial f_0}{\partial x}(X_0, t)X_1 + f_1(X_0, t), \quad X_1(0) = x_{01},$$

$$\dot{X}_2 = \frac{\partial f_0}{\partial x}(X_0, t)X_2 + \frac{1}{2}\left(\frac{\partial^2 f_0}{\partial x^2}(X_0, t)X_1\right)X_1$$

$$+ \frac{\partial f_1}{\partial x}(X_0, t)X_1 + f_2(X_0, t), \quad X_2(0) = x_{02},$$

etc. Thus, X_0 is determined as the solution of the nonlinear limiting problem, and we can obtain all later X_j's successively in terms of the solution of the variational matrix initial value problem

$$\dot{\Phi} = \frac{\partial f_0}{\partial x}(X_0, t)\Phi, \quad \Phi(0) = I$$

or, equivalently, the linear Volterra matrix integral equation $\Phi(t) = I + \int_0^t (\partial f_0/\partial x)(X_0(s), s)\Phi(s)ds$. Note that this integral equation could be solved by successive approximations on $0 \leq t \leq T$, or one could analogously define the "matrizant" $\Phi(t)$ using the resolvent kernel [cf. Cochran (1972)]. Specifically, note that the jth term $X_j(t)$ in the asymptotic expansion of the solution $X(t, \epsilon)$ will satisfy a linear vector system

$$\dot{X}_j = \frac{\partial f_0}{\partial x}(X_0, t)X_j + g_{j-1}(X_0, X_1, \cdots, X_{j-1}, t),$$

where g_{j-1} is a known smooth function of preceding coefficients X_l, $l < j$, and t. Variation of parameters, then, uniquely determines each such X_j, in turn, as

$$X_j(t) = \Phi(t)x_{0j} + \Phi(t)\int_0^t \Phi^{-1}(s)g_{j-1}(X_0(s), \cdots, X_{j-1}(s), s)ds$$

[since $\Phi(t)\Phi^{-1}(s)$ is the Green's function for the variational problem].

To show that the formal power series we have obtained is asymptotic, we define the remainder $\epsilon^N R_N$ after N terms by

$$\epsilon^N R_N(t, \epsilon) = X(t, \epsilon) - \sum_{j=0}^{N-1} \epsilon^j X_j(t)$$

and proceed to show that R_N is bounded on the fixed interval $0 \le t \le T$ for ϵ sufficiently small. The differential system implies that

$$\epsilon^N \dot{R}_N(t,\epsilon) = f(X_0 + \cdots + \epsilon^{N-1}X_{N-1} + \epsilon^N R_N, t, \epsilon) - \sum_{j=0}^{N-1} \epsilon^j \dot{X}_j(t)$$

$$= f(X_0 + \cdots + \epsilon^N R_N, t, \epsilon) - f_0(X_0, t)$$

$$- \sum_{j=1}^{N-1} \left(\frac{\partial f_0}{\partial x}(X_0, t)X_j + g_{j-1}(X_0, \cdots, X_{j-1}, t) \right) \epsilon^j.$$

Using the Taylor series for f, we can divide both sides by ϵ^N to obtain a nearly linear differential equation for R_N of the form

$$\dot{R}_N(t,\epsilon) = \frac{\partial f_0}{\partial x}(X_0, t)R_N + \epsilon h_N(X_0, \cdots, X_{N-1}, \ R_N, t, \epsilon)$$

with h_N bounded. Likewise,

$$R_N(0,\epsilon) = \frac{1}{\epsilon^N} \left(x_0(\epsilon) - \sum_{j=0}^{N-1} x_{0j}\epsilon^j \right) \equiv k(\epsilon)$$

is also bounded. Integrating this initial value problem finally implies the integral equation

$$R_N(t,\epsilon) = \Phi(t)k(\epsilon) + \epsilon\Phi(t) \int_0^t \Phi^{-1}(s)h_N(X_0(s), \cdots, R_N(s,\epsilon), s, \epsilon)ds$$

using the matrizant Φ defined above. For ϵ small, it is easy to use successive approximations to show that there is a unique bounded solution $R_N(t,\epsilon)$ on $0 \le t \le T$. This, of course, implies that the formal power series obtained for $x(t,\epsilon)$ converges asymptotically for all sufficiently small $|\epsilon|$ values. We note that the series for $X(t,\epsilon)$ would need to be terminated after a finite number of terms if the smoothness of f (and thereby X) were limited. The series even converges on $0 \le t \le T$ for small $|\epsilon|$ values. One would generally, however, expect the asymptotic validity of the expansion to break down as $t \to \infty$, even if $X_0(t)$ were defined and smooth throughout $t \ge 0$.

Exercise

[Compare Greenberg (1978)]

Consider the initial value problem

$$\dot{x} + x + \epsilon x^2 = 0, \qquad x(0) = \cos \epsilon.$$

a. Find the exact solution on $t \geq 0$.

b. Find the asymptotic solution to within $O(\epsilon^3)$ by a regular perturbation procedure.

c. Show that the formal power series expansion can be obtained by expanding the exact solution for ϵ small. Note that no complications occur at $t = \infty$, since the solution and the terms of the series expansion both decay exponentially to zero.

Chapter 2
Singularly Perturbed Initial Value Problems

A. A Nonlinear Problem from Enzyme Kinetics

Readers should refer to Murray (1977) and to earlier chemical engineering literature [especially Bowen et al. (1963) and Heinekin et al. (1967)] for experts' explanations of the significance of the pseudo-steady-state hypothesis in biochemistry. The theory of Michaelis and Menton (1913) and Briggs and Haldane (1925) concerns a substrate S being converted irreversibly by a single enzyme E into a product P. There is also an intermediate substrate–enzyme complex SE. Since the back reaction is negligible, we shall systematically write

$$S + E \underset{k_{-1}}{\overset{k_1}{\rightleftharpoons}} SE \overset{k_2}{\longrightarrow} P + E.$$

Using the law of mass action, we shall take the rates of reactions to be proportional to the concentrations of the reactants. Introducing s, e, c and p to denote the respective concentrations of S, E, SE, and P, we thereby obtain the nonlinear autonomous differential system

$$\begin{cases} \dfrac{ds}{dt} = -k_1 se + k_{-1}c, \\[2mm] \dfrac{de}{dt} = -k_1 se + (k_{-1} + k_2)c, \\[2mm] \dfrac{dc}{dt} = k_1 se - (k_{-1} + k_2)c, \\[2mm] \dfrac{dp}{dt} = k_2 c \end{cases}$$

subject to the initial conditions $s(0) = s_0 > 0$, $e(0) = e_0 > 0$, $c(0) = 0$, and $p(0) = 0$. Since $d(e+c)/dt = 0$ and $d(s+c+p)/dt = 0$,

$$e(t) = e_0 - c(t)$$

and

$$p(t) = s_0 - s(t) - c(t),$$

and there remains a nonlinear initial value problem for the concentrations s and c, viz.,

$$\begin{cases} \dfrac{ds}{dt} = -k_1 e_0 s + (k_1 s + k_{-1})c, & s(0) = s_0, \\[2mm] \dfrac{dc}{dt} = k_1 e_0 s - (k_1 s + k_{-1} + k_2)c, & c(0) = 0. \end{cases}$$

We note that such redundancy in the original differential system often occurs in chemical kinetics, circuit analysis, and other fields, due to constraints between variables which result from physical conservation or balance laws. Here, the resulting initial value problem for the two-dimensional system could be studied by describing all representative trajectories in the first quadrant of the s–c phase plane. Biochemists often explain the Michaelis–Menten kinetics less mathematically as follows: They expect $de/dt \approx 0$. Thus, $dc/dt \approx 0$, so $c \approx k_1 e_0 s / (k_1 s + k_{-1} + k_2)$ and $ds/dt \approx -k_2 e_0 s / (K + s)$ for $K \equiv (k_{-1} + k_2)/k_1$. Note that this pseudo-steady-state approach sets one derivative in this system equal to zero, but retains the other. It is used extensively, though it is not always valid [as at $t = 0$, where $c(0) = 0$ while $s(0) > 0$].

To better understand the true solution behavior, let us nondimensionalize variables by setting

$$\bar{t} = k_1 e_0 t, \quad \lambda = \frac{k_2}{k_1 s_0}, \quad \kappa = \frac{k_{-1} + k_2}{k_1 s_0},$$

$$x(\bar{t}) = \frac{s(t)}{s_0}, \quad y(\bar{t}) = \frac{c(t)}{e_0}, \quad \text{and } \epsilon = \frac{e_0}{s_0}.$$

Typically, $\epsilon \sim 10^{-6}$. Omitting the bar on t, then, we obtain the singularly perturbed initial value problem

$$\begin{cases} \dfrac{dx}{dt} = -x + (x + \kappa - \lambda)y, & x(0) = 1, \\[2mm] \epsilon \dfrac{dy}{dt} = x - (x + \kappa)y, & y(0) = 0. \end{cases}$$

[This long-popular scaling has recently been improved upon by Segel and Slemrod (1989) using the parameter $\mu = e_o/s_0(\kappa + 1) = \epsilon/(\kappa + 1)$ which is small with ϵ and also when κ is large. The resulting asymptotic solution for μ small will thereby be valid for a larger range of physical parameter values.] Assuming that κ and λ are bounded positive constants, the corresponding reduced differential-algebraic initial value problem

$$\begin{cases} \dfrac{dX_0}{dt} = -X_0 + (X_0 + \kappa - \lambda)Y_0, & X_0(0) = 1, \\[2mm] 0 = X_0 - (X_0 + \kappa)Y_0 \end{cases}$$

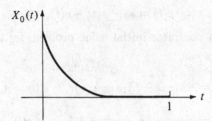

Fig. 2.1. Substrate concentration.

has an outer solution determined by the algebraic equation

$$Y_0 = \frac{X_0}{X_0 + \kappa}$$

and the reduced-order initial value problem

$$\frac{dX_0}{dt} = \frac{-\lambda X_0}{X_0 + \kappa}, \qquad X_0(0) = 1.$$

They together correspond to the pseudo-steady-state hypothesis of Michaelis and Menton. Note that $X_0(t)$ will decrease monotonically toward the rest point at the origin. Further, since $(-\lambda/\kappa)X_0 \leq dX_0/dt \leq -\lambda X_0/(1+\kappa) \leq 0$, $X_0(t)$ satisfies

$$0 \leq e^{-\lambda t/\kappa} \leq X_0(t) \leq e^{-\lambda t/(1+\kappa)} \leq 1.$$

[This estimate may actually be more useful than the implicit solution $X_0 + \kappa \ln X_0 = 1 - \lambda t$. However, a straightforward numerical integration on any bounded interval and the exponential bounds together simply determine the behavior of $X_0(t)$ throughout $t \geq 0$.] Pictorially, we will have Figure 2.1. Moreover, Y_0 will behave much like X_0. Because $Y_0(0) = 1/(1+\kappa) \neq y(0) = 0$, the limiting outer solution Y_0 cannot describe the fast variable y near $t = 0$. We anticipate having a solution y as shown in Figure 2.2.

The need for an initial layer correction in y suggests that we seek an asymptotic solution of the form

$$\begin{cases} x(t,\epsilon) = X(t,\epsilon) + \epsilon\xi(\tau,\epsilon), \\ y(t,\epsilon) = Y(t,\epsilon) + \eta(\tau,\epsilon) \end{cases}$$

with an outer solution $\binom{X}{Y}$ and an initial layer correction $\binom{\epsilon\xi}{\eta}$ which tends to zero as the stretched variable

$$\tau = \frac{t}{\epsilon}$$

tends to infinity. [The selection of τ can be further motivated by noting the size of $dy(0,\epsilon)/dt = 1/\epsilon$ as $\epsilon \to 0$.]

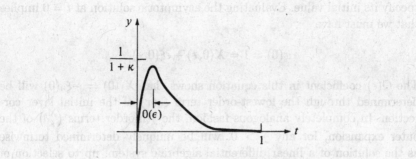

Fig. 2.2. Concentration of complex.

The outer solution must clearly be a smooth solution of our differential system

$$\begin{cases} \dfrac{dX}{dt} = -X + (X + \kappa - \lambda)Y, \\ \epsilon\dfrac{dY}{dt} = X - (X + \kappa)Y \end{cases}$$

which has an asymptotic power series expansion

$$\begin{pmatrix} X(t,\epsilon) \\ Y(t,\epsilon) \end{pmatrix} \sim \sum_{j=0}^{\infty} \begin{pmatrix} X_j(t) \\ Y_j(t) \end{pmatrix} \epsilon^j.$$

Thus, $\begin{pmatrix} X_0 \\ Y_0 \end{pmatrix}$ will satisfy the reduced system and later terms $\begin{pmatrix} X_j \\ Y_j \end{pmatrix}$ will satisfy linearized variational differential-algebraic systems obtained by successively equating coefficients of higher powers of ϵ to zero. From the coefficient of ϵ^1, we obtain

$$\begin{cases} \dfrac{dX_1}{dt} = -X_1 + X_1 Y_0 + (X_0 + \kappa - \lambda)Y_1, \\ \dfrac{dY_0}{dt} = X_1 - X_1 Y_0 - (X_0 + \kappa)Y_1. \end{cases}$$

Thus,

$$Y_1 = \frac{1}{X_0 + \kappa}\left((1 - Y_0)X_1 - \frac{dY_0}{dt} \right)$$

and there remains the linear differential equation

$$\frac{dX_1}{dt} = \frac{-\lambda\kappa}{(X_0 + \kappa)^2}X_1 - \frac{(X_0 + \kappa - \lambda)}{X_0 + \kappa}\frac{dY_0}{dt}.$$

(Because X_0 and dY_0/dt decay exponentially as $t \to \infty$ and the homogeneous equation has a decaying solution, the same will be true for X_1 and thereby Y_1.) An awkward formula for $X_1(t)$ is, indeed, provided by using the obvious integrating factor. To determine $X_1(t)$ uniquely, we need to specify its initial value. Evaluating the asymptotic solution at $t = 0$ implies that we must have

$$x(0) = 1 = X(0, \epsilon) + \epsilon\xi(0, \epsilon).$$

The $O(\epsilon)$ coefficient in this equation shows that $X_1(0) = -\xi_0(0)$ will be determined through the lowest-order term $\xi_0(\tau)$ of the initial layer correction. In completely analogous fashion, the jth-order terms $\binom{X_j}{Y_j}$ of the outer expansion, for any $j > 0$, will be uniquely determined termwise as the solution of a linear differential-algebraic system, up to selection of $X_j(0) = -\xi_{j-1}(0)$, which is specified by the $(j-1)$st-order term $\xi_{j-1}(\tau)$ of the initial layer correction. The pattern to formally obtain all successive terms of the asymptotic solution is already clear. One first solves the reduced problem to uniquely obtain $\binom{X_0(t)}{Y_0(t)}$. Then, one obtains (as we will next show) the leading terms $\binom{\xi_0(\tau)}{\eta_0(\tau)}$ of the decaying initial layer correction, followed successively by the $O(\epsilon)$ terms of the outer expansion and the $O(\epsilon)$ terms $\binom{\xi_1(\tau)}{\eta_1(\tau)}$ of the initial layer correction, etc.

Upon differentiating the assumed form of the asymptotic solution, we obtain

$$\frac{dx}{dt} = \frac{dX}{dt} + \epsilon\frac{d\xi}{d\tau}\frac{d\tau}{dt} = \frac{dX}{dt} + \frac{d\xi}{d\tau},$$

so

$$\frac{d\xi}{d\tau} = \frac{dx}{dt} - \frac{dX}{dt} = -(X + \epsilon\xi) + (X + \epsilon\xi + \kappa - \lambda)(Y + \eta)$$
$$- [-X + (X + \kappa - \lambda)Y] = (X + \kappa - \lambda)\eta + \epsilon(-1 + Y + \eta)\xi.$$

An equation for $d\eta/d\tau = \epsilon(dy/dt) - \epsilon(dY/dt)$ follows analogously. Thus, $\binom{\xi}{\eta}$ must satisfy the resulting nearly linear system

$$\begin{cases} \dfrac{d\xi}{d\tau} = (X + \kappa - \lambda)\eta + \epsilon(-1 + Y + \eta)\xi, \\ \dfrac{d\eta}{d\tau} = -(X + \kappa)\eta + \epsilon(1 - Y - \eta)\xi \end{cases}$$

through an asymptotic power series expansion

$$\binom{\xi(\tau, \epsilon)}{\eta(\tau, \epsilon)} \sim \sum_{j=0}^{\infty} \binom{\xi_j(\tau)}{\eta_j(\tau)}\epsilon^j$$

whose terms all decay to zero as $\tau \to \infty$. To obtain differential equations for successive terms, we expand the products $X(\epsilon\tau, \epsilon)\eta(\tau, \epsilon)$, $Y(\epsilon\tau, \epsilon)\xi(\tau, \epsilon)$, and $\eta(\tau, \epsilon)\xi(\tau, \epsilon)$ as power series in ϵ and equate coefficients of like powers of ϵ in the two differential equations in τ. Thus, when $\epsilon = 0$, we obtain the constant linear system

$$\begin{cases} \dfrac{d\xi_0}{d\tau} = (1 + \kappa - \lambda)\eta_0, \\[3mm] \dfrac{d\eta_0}{d\tau} = -(1 + \kappa)\eta_0 \end{cases}$$

since $X_0(0) = 1$, and from the coefficients of ϵ^1, we have the nonhomogeneous system

$$\begin{cases} \dfrac{d\xi_1}{d\tau} = (1 + \kappa - \lambda)\eta_1 + [\tau X_0'(0) + X_1(0)]\eta_0 + [-1 + Y_0(0) + \eta_0]\xi_0, \\[3mm] \dfrac{d\eta_1}{d\tau} = -(1 + \kappa)\eta_1 - [\tau X_0'(0) + X_1(0)]\eta_0 + [1 - Y_0(0) - \eta_0]\xi_0. \end{cases}$$

Analogous linear systems for later terms $\binom{\xi_j}{\eta_j}$, $j > 0$, follow in a straightforward manner. Their nonhomogeneous terms are completely specified by functions already known. To determine decaying solutions for these systems, we will only need to specify initial values for the η_j's. Note, however, that the initial condition

$$y(0) = 0 = Y(0, \epsilon) + \eta(0, \epsilon)$$

implies that

$$\eta_0(0) = -Y_0(0) = \frac{-1}{1 + \kappa}$$

and $\eta_j(0) = -Y_j(0)$ for all $j \geq 0$. Thus, we uniquely determine

$$\eta_0(\tau) = -\left(\frac{1}{1 + \kappa}\right) e^{-(1+\kappa)\tau}$$

and thereby $d\xi_0/d\tau$. Since $\xi_0(\infty) = 0$, we must have

$$\xi_0(\tau) = -\int_\tau^\infty \frac{d\xi_0}{d\tau}(s)ds = \frac{(1 + \kappa - \lambda)}{(1 + \kappa)^2} e^{-(1+\kappa)\tau}.$$

This, moreover, specifies the initial value

$$X_1(0) = -\xi_0(0)$$

and allows us to completely determine the first-order outer expansion term $X_1(t)$ and, thereby, $Y_1(t)$. Since we will then know

$$\eta_1(0) = -Y_1(0),$$

we will next be able to integrate the linear equation to determine $\eta_1(\tau)$. This, in turn, provides us $\xi_1(\tau) = -\int_\tau^\infty (d\xi_1/d\tau)(s)ds$ and thereby the initial value $X_2(0) = -\xi_1(0)$ needed to completely specify the second-order outer expansion terms. In practical situations, we note that only a few higher-order terms are typically used. The asymptotic validity of our approach will be shown below for much more general nonlinear vector systems, based on the work of Tikhonov and Levinson.

In the biochemical context, the concentrations of the enzyme e and the complex c change rapidly in the thin initial layer. Indeed, the rapid changes typically occur before any experimental measurements can get underway. Prime interest lies in the rates of reactions. Using the pseudo-steady-state hypothesis, one might take $dX_0(0)/dt = -\lambda/(1+\kappa)$ instead of the correct $dx(0,0)/dt = -1$. The error in approximating the asymptotically unbounded $dy(0,\epsilon)/dt = 1/\epsilon$ by $dY_0(0)/dt = -\lambda\kappa/(1+\kappa)^3$ is much less acceptable. Away from $t = 0$, however, the asymptotic arguments we will develop justify the Michaelis–Menton theory (at least when ϵ is small; otherwise, a careful numerical integration is called for).

Exercises

1. Consider the linear system

    $$\dot{x} = ax + by,$$
    $$\epsilon\dot{y} = cx + dy$$

    of two linear constant coefficient scalar equations on $t \geq 0$, with $d < 0$, subject to prescribed initial values for $x(0)$ and $y(0)$. Note that the problem might be rewritten in system form as $\dot{z} = \mathcal{A}(\epsilon)z$ for $z = \binom{x}{y}$ and

    $$\mathcal{A} = \begin{pmatrix} a & b \\ c/\epsilon & d/\epsilon \end{pmatrix}.$$

    The asymptotic behavior of the corresponding unique solution $z(t,\epsilon) = e^{\mathcal{A}(\epsilon)t}z(0)$ will be determined.

    a. Determine the asymptotic size of the eigenvalues $\lambda(\epsilon)$ and corresponding eigenvectors $p(\epsilon)$ of $\mathcal{A}$, and use them to determine two linearly independent solutions of the differential system in the form $z(t) = e^{\lambda(\epsilon)t}p(\epsilon)$.

    b. Take a linear combination of these solutions to satisfy the initial conditions.

    c. Show that the solution $z(t)$ obtained implies an asymptotic solution of the form
    $$x(t,\epsilon) = X_0(t) + O(\epsilon),$$
    $$y(t,\epsilon) = Y_0(t) + \eta_0(\tau) + O(\epsilon)$$

    on any bounded t interval $[O,T]$ where $\binom{X_0}{Y_0}$ satisfies the reduced problem and $\eta_0 \to 0$ as $\tau \equiv t/\epsilon \to \infty$.

2. Show that the solution to the scalar time-varying initial value problem

    $$\epsilon\dot{y} = a(t)y + b(t), \qquad y(0) \text{ given}$$

tends to $-b(t)/a(t)$ as $\epsilon \to 0$ for $t > 0$ provided $a(t) \leq -\delta < 0$ and b/a is continuously differentiable on $t \geq 0$. {Hint: Write down the exact solution and integrate by parts, noting that $\exp[(1/\epsilon) \int_s^t a(r)dr] \leq e^{-\delta(t-s)/\epsilon}$ for $t \geq s$.}

B. The Solution of Linear Systems Using Transformation Methods

Consider the homogeneous linear differential system

$$\begin{cases} \dot{x} = A(t)x + B(t)y, \\ \epsilon\dot{y} = C(t)x + D(t)y, \end{cases}$$

where x and y are, respectively, m- and n-dimensional vectors, where the matrices A, B, C, and D are smooth functions of t, and where $D(t)$ is *stable* (i.e., its eigenvalues λ all lie strictly in the left half of the complex plane with Re $\lambda(t) \leq -\sigma$ for some $\sigma > 0$). Suppose we wish to solve the system on a fixed bounded interval, say, $0 \leq t \leq 1$, with $x(0)$ and $y(0)$ prescribed. We might anticipate (on the basis of the scalar problem which we solved explicitly as an exercise) that the limiting solution for $t > 0$ should satisfy the reduced (differential-algebraic) initial value problem

$$\begin{cases} \dot{X}_0 = A(t)X_0 + B(t)Y_0, \quad X_0(0) = x(0), \\ 0 = C(t)X_0 + D(t)Y_0. \end{cases}$$

Because the matrix D is nonsingular, any such limit would yield

$$Y_0(t) = -D^{-1}(t)C(t)X_0(t),$$

where X_0 satisfies the mth-order initial value problem

$$\dot{X}_0 = (A(t) - B(t)D^{-1}(t)C(t))X_0, \quad X_0(0) = x(0).$$

We note that this reduced order system is not stiff, compared to the original system, so it should be relatively easy to integrate numerically. We also observe that the state matrix $A - BD^{-1}C$ is the Schur complement of D [cf. Cottle (1974)]. [It arises, for example, when we wish to solve the block matrix equation $\begin{pmatrix} A & B \\ C & D \end{pmatrix} \begin{pmatrix} \alpha \\ \beta \end{pmatrix} = \begin{pmatrix} f \\ g \end{pmatrix}$ by "tearing." If D is nonsingular, setting $\beta = -D^{-1}C\alpha + D^{-1}g$ reduces this linear algebra problem to solving the system $(A - BD^{-1}C)\alpha = f - BD^{-1}g$.] Unless $Y_0(0) = -D^{-1}(0)C(0)x(0)$ happens to equal $y(0)$, the vector y cannot be well represented by Y_0 near $t = 0$. For x and y bounded, we can expect $\dot{y} = O(1/\epsilon)$ and $\dot{x} = O(1)$ as $\epsilon \to 0$. Thus, it is natural to call x the slow and y the (at least potentially) fast solution vector. Indeed, we might already expect y to have an $O(\epsilon)$

thick t-interval of nonuniform convergence near $t = 0$ and an asymptotic solution of the form

$$\begin{cases} x(t, \epsilon) = X(t, \epsilon) + \epsilon \xi(\tau, \epsilon), \\ y(t, \epsilon) = Y(t, \epsilon) + \eta(\tau, \epsilon), \end{cases}$$

where ξ and $\eta \to 0$ as $\tau = t/\epsilon \to \infty$.

We will determine the solution structure for this linear problem by actually decoupling the fast and slow parts of our solution through changes of variables which will (ultimately) block-diagonalize the coefficient matrix of our linear system. [First-time readers are warned that they should not allow themselves to get bogged down in the algebraic and analytic details which follow. The approach presented below has much value for many linear problems (including our later topic of singularly perturbed boundary value problems (cf. Section 3C).] For an historical review of such Riccati transformation techniques, see Smith (1987). Details of proof are left as exercises. Alternatively, proofs also follow from the nonlinear Tikhonov–Levinson theory (cf. Sections 2D and 2E).

Let us, first, attempt to define a purely fast variable v by splitting off the slow part of y. An important motivation for splitting the solution into (purely) slow- and fast-varying parts is the natural inclination to perform numerical integrations using a relatively fine mesh on only the fast-varying parts. As an ansatz, let us set

$$v(t) = y(t) + L(t, \epsilon)x(t)$$

or, equivalently,

$$y(t) = v(t) - L(t, \epsilon)x(t)$$

for an appropriate $n \times m$ matrix L. Since

$$\epsilon \dot{v} = \epsilon \dot{y} + \epsilon L\dot{x} + \epsilon \dot{L}x = (\epsilon \dot{L} + \epsilon LA + C)x + (D + \epsilon LB)y$$
$$= (D + \epsilon LB)v + [\epsilon \dot{L} + \epsilon LA + C - (D + \epsilon LB)L]x,$$

we will obtain the "purely fast" and stable (as $\epsilon \to 0$) equation

$$\epsilon \dot{v} = (D + \epsilon LB)v,$$

provided L is selected as a bounded solution of the singularly perturbed matrix Riccati equation

$$\epsilon \dot{L} = DL - \epsilon LA + \epsilon LBL - C.$$

Note that this matrix differential equation reduces to a nonsingular linear algebraic system as $\epsilon \to 0$. We shall determine a unique smooth outer solution of this equation which has an asymptotic series expansion

$$L(t, \epsilon) \sim \sum_{j=0}^{\infty} L_j(t)\epsilon^j$$

(presuming infinite differentiability of the original coefficient matrices A, B, C, and D). Equating coefficients successively, then, we first have

$$DL_0 - C = 0$$

and, then, for each $j \geq 1$,

$$DL_j = \dot{L}_{j-1} + L_{j-1}A - \sum_{k=0}^{j-1} L_k BL_{j-1-k}.$$

Since D is invertible, this uniquely determines all matrix coefficients L_j successively. [With less smoothness or less patience, we would need to terminate the series after a finite number of terms, obtaining an approximate decoupling to a corresponding order, say $O(\epsilon^N)$.] That the resulting formal expansion for $L(t, \epsilon)$ is asymptotically correct will follow from our general nonlinear theory. The critical hypothesis is the stability of the matrix $D(t)$ throughout $0 \leq t \leq 1$.

Exercises

1. Consider the constant coefficient system

$$\begin{cases} \dot{x} = x - 2y, \\ \epsilon\dot{y} = 3x - 4y \end{cases}$$

of two scalar equations. Solve the initial value problem on $t \geq 0$ by converting the system to the triangular form

$$\begin{cases} \dot{x} = \alpha x + \beta v, \\ \epsilon\dot{v} = \gamma v \end{cases}$$

(for appropriate scalars α, β, and γ) using a change of variables $v = y + lx$.

2.

   a. Show that the matrix initial value problem

$$\epsilon\dot{Z} = D(t)Z, Z(0) = I$$

   has a unique *fundamental matrix* solution with boundary layer behavior near $t = 0$ when $D(t)$ is a stable $n \times n$ matrix. *Idea:* When D is constant, $Z(t, \epsilon) = e^{Dt/\epsilon}$ has the right behavior since Re $\lambda(D) < -\sigma < 0$ for all eigenvalues λ of D implies that $Z(t, \epsilon) = O(e^{-\sigma t/\epsilon})$.

[For a discussion of the matrix exponential and much other background for this exercise, see Coddington and Levinson (1955), Coppel (1965), and Brockett (1970).] More generally, for a time-varying $D(t)$, Z will satisfy the linear matrix integral equation

$$Z(t,\epsilon) = I + \frac{1}{\epsilon} \int_0^t D(s)Z(s,\epsilon)ds.$$

It can be uniquely solved on, say, $0 \leq t \leq 1$, via a resolvent kernel or, equivalently, successive approximations, i.e., by setting

$$\begin{cases} Z^o(t,\epsilon) = I, \\ Z^j(t,\epsilon) = I + \frac{1}{\epsilon} \int_0^t D(s)Z^{j-1}(s,\epsilon)ds, \quad j \geq 1, \\ Z(t,\epsilon) = \lim_{j\to\infty} Z^j(t,\epsilon). \end{cases}$$

Show that each column z of Z separately satisfies $z'\dot{z} = \frac{1}{\epsilon}z'D(t)z$, where the prime denotes transposition. Using the vector inner product norm $\|z\| = \sqrt{z'z}$, it follows that $\frac{d}{dt}\|z\| \leq -\frac{\sigma}{\epsilon}\|z\|$ since Re $\lambda(D(t)) < -\sigma < 0$ for all eigenvalues $\lambda(t)$. Thus, the matrizant Z satisfies $Z(t,\epsilon) = O(\epsilon^{-\sigma t/\epsilon})$, i.e., the elements of Z are all "of boundary layer type" with a trivial outer limit. We note that it may, indeed, be more convenient or suggestive to alternatively express the fundamental solution $Z(t,\epsilon)$ as a matrix $\tilde{Z}(\tau,\epsilon)$ which tends to zero as the stretched variable $\tau = t/\epsilon \to \infty$.

b. Using variation of parameters on the linear part of the matrix Riccati system

$$\epsilon\dot{L} - DL = f \equiv -C - \epsilon L(A - BL)$$

and using the fundamental matrix $Z(t,\epsilon)$ obtained above, show that L will satisfy

$$L(t,\epsilon) = Z(t,\epsilon)\left(L(0,\epsilon) - \frac{1}{\epsilon}\int_0^t Z^{-1}(s,\epsilon)C(s)ds\right.$$

$$\left. - \int_0^t Z^{-1}(s,\epsilon)L(s,\epsilon)[A(s) - B(s)L(s,\epsilon)]ds\right).$$

Its solution can then be obtained by the successive approximations scheme:

$$\begin{cases} L^o(t,\epsilon) = Z(t,\epsilon)\left(L(0,\epsilon) - \frac{1}{\epsilon}\int\limits_0^t Z^{-1}(s,\epsilon)C(s)ds \right), \\[2em] L^j(t,\epsilon) = L^o(t,\epsilon) - Z(t,\epsilon)\int\limits_0^t Z^{-1}(s,\epsilon)L^{j-1}(s,\epsilon)[A(s) \\[1em] \qquad\qquad\qquad - B(s)L^{j-1}(s,\epsilon)]ds, \ \ j \geq 1, \\[1em] L(t,\epsilon) \equiv \lim\limits_{j\to\infty} L^j(t,\epsilon). \end{cases}$$

Note: This procedure works for any bounded initial value $L(0,\epsilon)$. In d, we will show that L will be smooth for the "right" initial matrix $L(0,\epsilon)$.

c. Since $ZZ^{-1} = I$, show that Z^{-1} will satisfy the differential equation $\epsilon(Z^{-1})^\bullet = -Z^{-1}D(t)$. [Because $-D(t)$ has its eigenvalues in the right half-plane, it readily follows that Z^{-1} grows exponentially as t increases. Thus, it might be convenient to instead represent Z^{-1} on $0 \leq t \leq 1$ as a terminal boundary layer function of $\sigma = (1-t)/\epsilon$ which decays to zero as $\sigma \to \infty$.]

d. For any smooth f, use integration by parts to show that

$$\frac{1}{\epsilon}Z(t,\epsilon)\int\limits_0^t Z^{-1}(s,\epsilon)f(s)ds = -\, D^{-1}(t)f(t) + Z(t,\epsilon)D^{-1}(0)f(0)$$

$$+ Z(t,\epsilon)\int\limits_0^t Z^{-1}(s,\epsilon)[D^{-1}(s)f(s)]^\bullet ds.$$

Repeat the process again to show that the last integral can be replaced by

$$-\epsilon D^{-1}(t)[D^{-1}(t)f(t)]^\bullet + \epsilon Z(t,\epsilon)D^{-1}(0)[D^{-1}(t)f(t)]^\bullet\big|_{t=0}$$

$$+ \epsilon Z(t,\epsilon)\int\limits_0^t Z^{-1}(s,\epsilon)\{D^{-1}(s)[D^{-1}(s)f(s)]^\bullet\}^\bullet ds.$$

Thus,

$$\frac{1}{\epsilon} Z(t,\epsilon) \int_0^t Z^{-1}(s,\epsilon) f(s) ds = -D^{-1}(t) \left\{ f(t) + \epsilon [D^{-1}(t) f(t)]^\bullet \right\}$$

$$+ Z(t,\epsilon) D^{-1}(0) \left\{ f(0) + \epsilon [D^{-1}(t) f(t)]^\bullet \big|_{t=0} \right\}$$

$$+ \epsilon Z(t,\epsilon) \int_0^t Z^{-1}(s,\epsilon) \left\{ D^{-1}(s) [D^{-1}(s) f(s)]^\bullet \right\}^\bullet ds.$$

Use the preceding result for $f = -C - \epsilon L(A - BL)$ and the integral equation for L to show that

$$L(t,\epsilon) = D^{-1}(t) C(t) + Z(t,\epsilon)[L(0,\epsilon) - D^{-1}(0) C(0)] + O(\epsilon)$$

and, because Z is fast decaying, that the smooth solution $L(t,\epsilon)$ of the matrix Riccati equation will satisfy $L(t,\epsilon) = D^{-1}(t) C(t) + O(\epsilon)$. It is straightforward to show that nonsmooth solutions of the Riccati equation featuring nonuniform convergence near $t = 0$ will occur, except for the special initial value $L(0,\epsilon)$ corresponding to the formally constructed smooth solution $L(t,\epsilon)$. {That a matrix $L(0,\epsilon)$ exists having the constructed asymptotic series follows from the Borel–Ritt Theorem [cf. Wasow (1965)].}

Having, at least formally, determined a smooth matrix solution $L(t,\epsilon)$ of the Riccati equation, let us seek a solution v of the resulting purely fast system

$$\epsilon \dot{v} = (D(t) + \epsilon L(t,\epsilon) B(t)) v$$

subject to the initial condition

$$v(0) = y(0) + L(0,\epsilon) x(0).$$

Since D is nonsingular and the equation for v is homogeneous, any attempt to find an outer expansion

$$V(t,\epsilon) \sim \sum_{j \geq 0} V_j(t) \epsilon^j$$

for v will result in the trivial series (i.e., $V_j \equiv 0$ for every j). Indeed, the stability of D implies that it is more natural to directly seek a solution consisting of only an initial layer expansion

$$v(\tau,\epsilon) \sim \sum_{j=0}^\infty v_j(\tau) \epsilon^j$$

whose terms all tend to zero as $\tau = t/\epsilon$ tends to infinity. Substituting $t = \epsilon\tau$ and expanding both sides of the differential system

$$\frac{dv}{d\tau} = [D(\epsilon\tau) + \epsilon L(\epsilon\tau, \epsilon)B(\epsilon\tau)]\,v$$

as a power series in ϵ, we successively obtain systems for the v_j's in the form

$$\frac{dv_j}{d\tau} = D(0)v_j + \alpha_{j-1}(\tau),$$

where $\alpha_{-1}(\tau) \equiv 0$ and (an inductive argument shows that) later α_{j-1}'s become successively known in terms of preceding coefficients as exponentially decaying vectors which tend to zero as $\tau \to \infty$. Thus, using the matrix exponential $e^{D(0)\tau}$ and variation of parameters, we obtain

$$v_j(\tau) = e^{D(0)\tau}v_j(0) + \int_0^\tau e^{D(0)(\tau-s)}\alpha_{j-1}(s)ds.$$

The initial condition implies that $v_0(0) = y(0) + D^{-1}(0)C(0)x(0)$ and $v_j(0) = L_j(0)x(0)$ for each $j > 0$. Like the matrix $e^{D(0)\tau}$, each vector v_j is easily shown to satisfy $v_j(\tau) = O(e^{-\sigma\tau})$ under our stability hypothesis. The initial layer expansion $v(\tau, \epsilon)$, then, matches the trivial outer solution $V(t, \epsilon) \equiv 0$ as $\tau \to \infty$. We note that the solution $v(\tau, \epsilon)$, which we have formally obtained, corresponds to the solution of the linear integral equation

$$v(\tau, \epsilon) = e^{D(0)\tau}[y(0) + L(0, \epsilon)x(0)] + \int_0^\tau e^{D(0)(\tau-s)}[D(\epsilon s) - D(0)$$

$$+ \epsilon L(\epsilon s, \epsilon)B(\epsilon s)]v(s, \epsilon)ds.$$

It could also be obtained by using a resolvent kernel or various approximate iteration methods. Note that the integrand is $O(\epsilon)$, so a unique decaying solution v is readily obtained.

Having found L and v, the predominantly slow vector x will satisfy the nonhomogeneous system

$$\dot{x} = (A - BL)x + Bv$$

which is forced by the fast-decaying function $B(t)v(t/\epsilon, \epsilon)$. This introduces a small fast component into the primarily slow variable x, so it is now convenient to introduce a "purely slow" vector u through a new transformation

$$u = x + \epsilon Hv \qquad (\text{or } x = u - \epsilon Hv)$$

obtained by using an appropriate $m \times n$ matrix H. [Note that this implies the alternative representation $u = (I + \epsilon HL)x + \epsilon Hy$ in terms of the original variables.] Differentiating, we have $\dot{u} = \dot{x} + \epsilon \dot{H}v + \epsilon H\dot{v} = (A - BL)(u - \epsilon Hv) + [\epsilon\dot{H} + H(D + \epsilon LB) + B]v$, so u will satisfy the purely slow system

$$\dot{u} = (A - BL)u,$$

provided H satisfies the singularly perturbed linear matrix system

$$\epsilon \dot{H} = -H(D + \epsilon LB) + \epsilon(A - BL)H - B$$

(which reduces to a nonsingular linear algebraic equation when $\epsilon = 0$). We naturally seek a smooth solution $H(t, \epsilon)$ with an asymptotic series expansion

$$H(t, \epsilon) \sim \sum_{j=0}^{\infty} H_j(t)\epsilon^j.$$

Equating coefficients successively, we first require

$$H_0(t) = -B(t)D^{-1}(t)$$

and, then, for each $j > 0$,

$$H_j(t) = \left[-\dot{H}_{j-1} + AH_{j-1} - \sum_{k=0}^{j-1}(H_k L_{j-1-k}B + BL_{j-1-k}H_k) \right] D^{-1}.$$

To show that the formally obtained series corresponds to a smooth solution will be left as the following exercise for the reader.

Exercise

a.   Consider the linear matrix equation

$$\dot{Z} = M(t)Z + ZN(t) + P(t)$$

and let $\mathcal{A}(t)$ and $\mathcal{B}(t)$ be nonsingular solutions of the matrix systems $\dot{\mathcal{A}} = M(t)\mathcal{A}$ and $\dot{\mathcal{B}} = \mathcal{B}N(t)$, respectively. Verify that the general matrix solution Z will be given by

$$Z(t) = \mathcal{A}(t) \int^t \mathcal{A}^{-1}(s)P(s)\mathcal{B}^{-1}(s)ds\mathcal{B}(t).$$

b.   Introduce $\mathcal{M}(t, \epsilon)$ and $\mathcal{N}(\sigma, \epsilon)$ as nonsingular solutions of the linear systems

$$\frac{d\mathcal{M}}{dt} = (A(t) - B(t)L(t, \epsilon))\mathcal{M}$$

and

$$\frac{d\mathcal{N}}{d\sigma} = \mathcal{N}(D(1 - \epsilon\sigma) + \epsilon L(1 - \epsilon\sigma, \epsilon)B(1 - \epsilon\sigma)),$$

respectively. Since $D(1)$ is stable, $\mathcal{N}$ will have an asymptotic expansion in ϵ whose terms decay exponentially to zero as $\sigma \to \infty$. Thus, $\mathcal{N}((1 - t)/\epsilon)$ will feature boundary layer behavior near $t = 1$ and it will match the differential system's trivial outer expansion on $0 \leq t < 1$. Show that the linear matrix differential system

$$\epsilon \dot{h} = \epsilon(A - BL)h - h(D + \epsilon LB) - B$$

has the general solution

$$h(t, \epsilon) = \mathcal{M}(t, \epsilon)\left[\mathcal{M}^{-1}(1, \epsilon)h(1, \epsilon)\mathcal{N}^{-1}(0, \epsilon) \right.$$

$$\left. + \frac{1}{\epsilon}\int\limits_{t}^{1} \mathcal{M}^{-1}(s, \epsilon)B(s)\mathcal{N}^{-1}\left(\frac{1-s}{\epsilon}, \epsilon\right) ds \right]\mathcal{N}\left(\frac{1-t}{\epsilon}, \epsilon\right)$$

on $0 \leq t \leq 1$. Note that this solution will have the asymptotic form $h(t, \epsilon) = H(t, \epsilon) + \lambda(\sigma, \epsilon)$ for an outer expansion H and a terminal layer correction λ which decays to zero as $\sigma = (1 - t)/\epsilon \rightarrow \infty$. Also note that $H(t, \epsilon)$ is the unique smooth solution corresponding to a special terminal value $h(1, \epsilon)$ which agrees asymptotically with the series we formally generated.

Having determined the smooth solution $H(t, \epsilon)$ of the linear matrix system as a formal power series, it only remains to solve the purely slow linear system

$$\dot{u} = (A(t) - B(t)L(t, \epsilon))u$$

with the initial vector

$$u(0, \epsilon) = [x(0) + \epsilon H(0, \epsilon)v(0, \epsilon)]$$
$$= [I + \epsilon H(0, \epsilon)L(0, \epsilon)]x(0) + \epsilon H(0, \epsilon)y(0).$$

Its solution can readily be determined as a smooth outer expansion

$$U(t, \epsilon) \sim \sum_{j=0}^{\infty} U_j(t)\epsilon^j.$$

The leading term must satisfy the reduced problem

$$\dot{U}_0 = (A - BL_0)U_0 = (A - BD^{-1}C)U_0, \quad U_0(0) = x(0),$$

so uniqueness implies that

$$U_0(t) = X_0(t).$$

Succeeding terms will satisfy the nonhomogeneous initial value problems

$$\dot{U}_j = (A - BD^{-1}C)U_j - B\sum_{k=0}^{j-1} L_{j-k}U_k, \quad \text{with } U_j(0) \text{ as prescribed.}$$

To summarize, we have shown that the two-step change of variables

$$\begin{pmatrix} u \\ v \end{pmatrix} = \begin{pmatrix} I & \epsilon H \\ 0 & I \end{pmatrix} \begin{pmatrix} I & 0 \\ L & I \end{pmatrix} \begin{pmatrix} x \\ y \end{pmatrix} = \begin{pmatrix} I + \epsilon HL & \epsilon H \\ L & I \end{pmatrix} \begin{pmatrix} x \\ y \end{pmatrix},$$

for smooth matrices $L(t, \epsilon)$ and $H(t, \epsilon)$, converts our original initial value problem for x and y to a pair of decoupled problems for the purely slow system

$$\dot{u} = (A - BL)u$$

and the purely fast system

$$\epsilon \dot{v} = (D + \epsilon LB)v.$$

It also determines the initial values for u and v in terms of those for x and y. Note that

$$\begin{pmatrix} x \\ y \end{pmatrix} = \begin{pmatrix} I & 0 \\ -L & I \end{pmatrix} \begin{pmatrix} I & -\epsilon H \\ 0 & I \end{pmatrix} \begin{pmatrix} u \\ v \end{pmatrix} = \begin{pmatrix} I & -\epsilon H \\ -L & I + \epsilon LH \end{pmatrix} \begin{pmatrix} u \\ v \end{pmatrix},$$

where

$$u(t, \epsilon) = \mathcal{U}(t, \epsilon)u(0, \epsilon) \quad \text{and} \quad v(\tau, \epsilon) = \mathcal{V}(\tau, \epsilon)v(0, \epsilon)$$

could be obtained as formal power series or by using fundamental matrices $\mathcal{U}(t, \epsilon)$ and $\mathcal{V}(\tau, \epsilon)$ for the respective linear systems, normalized such that $\mathcal{U}(0, \epsilon) = I$ and $\mathcal{V}(0, \epsilon) = I$. It follows that the solution of the original problem has the anticipated form

$$x(t, \epsilon) = X(t, \epsilon) + \epsilon \xi(\tau, \epsilon),$$
$$y(t, \epsilon) = Y(t, \epsilon) + \eta(\tau, \epsilon)$$

with the outer solution

$$\begin{pmatrix} X(t, \epsilon) \\ Y(t, \epsilon) \end{pmatrix} = \begin{pmatrix} I \\ -L(t, \epsilon) \end{pmatrix} \mathcal{U}(t, \epsilon) \left\{ [I + \epsilon H(0, \epsilon)L(0, \epsilon)]x(0) + \epsilon H(0, \epsilon)y(0) \right\}$$

and the initial layer correction

$$\begin{pmatrix} \epsilon \xi(\tau, \epsilon) \\ \eta(\tau, \epsilon) \end{pmatrix} = \begin{pmatrix} -\epsilon H(\epsilon \tau, \epsilon) \\ I + \epsilon L(\epsilon \tau, \epsilon)H(\epsilon \tau, \epsilon) \end{pmatrix} \mathcal{V}(\tau, \epsilon)[y(0) + L(0, \epsilon)x(0)].$$

Exercises

1. Explicitly solve the initial value problem

$$\dot{x} = x - 2y,$$
$$\epsilon \dot{y} = 3x - 4y$$

on $0 \le t \le 1$ by completely decoupling the slow and fast dynamics.

2. Determine which initial vectors $x(0)$ correspond to bounded solutions of the two-dimensional linear system $\epsilon \dot{x} = Ax$ for $A = \begin{pmatrix} 2 & 3 \\ -1 & -2 \end{pmatrix}$.

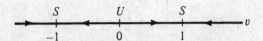

Fig. 2.3. Stable (and unstable) rest points $v = \pm 1$ (and $v = 0$).

C. Inner and Outer Solutions of Model Problems

(i)   Consider the nonlinear scalar initial value problem for

$$\epsilon \dot{y} = y - y^3$$

on $t \geq 0$. The reduced problem

$$Y_0 - Y_0^3 = 0$$

(and, indeed, also the full equation for $\epsilon > 0$) has the three constant solutions $Y_0(t) = 0, 1$, and -1. If $y(0)$ is not one of these three values, we will have $\dot{y}(0) = O(1/\epsilon)$, so we might hope to capture the rapid initial motion by introducing the stretched time

$$\tau = t/\epsilon$$

and seeking a solution $v(\tau, \epsilon)$ of the inner problem

$$\frac{dv}{d\tau} = v(1 - v)(1 + v), \quad v(0) = y(0)$$

on the semi-infinite interval $\tau \geq 0$ (i.e., for $t = 0^+$). Our singular perturbation problem has thereby been conveniently converted to a stability problem on $\tau \geq 0$. Note that the sign of $dv/d\tau$ implies that v will decrease (or, respectively, increase) monotonically to $v(\infty) = 1$ if $y(0) > 1$ [or $0 < y(0) < 1$] and that it will decrease (increase) monotonically to $v(\infty) = -1$ if $-1 < y(0) < 0$ [$y(0) < -1$]. Pictorially, we have the direction field shown in Figure 2.3 for increasing τ. The equation can be explicitly solved as a Riccati equation for v^2 or a separable equation for v to yield

$$v(\tau) = y(0)/\sqrt{y^2(0) + [1 - y^2(0)]e^{-2\tau}},$$

which we could rewrite in the form $v(\tau) = v(\infty) + \xi(\tau)$, where $\xi \to 0$ as $\tau \to \infty$. Then, matching implies that the limiting solution for $t > 0$ is $Y_0(t) = 1$ if $y(0) > 0$ and $Y_0(t) = -1$ if $y(0) < 0$. The trivial solution holds if $y(0) = 0$, but it is clearly unstable to small perturbations of the initial value. The true solution, therefore, converges nonuniformly to a unique constant limit as $\epsilon \to 0$ in an $O(\epsilon)$ thick initial layer, provided $y^2(0) \neq 0$ or 1.

(ii) A similar analysis applies for the two-dimensional system

$$\begin{cases} \dot{x} = yx, \\ \epsilon\dot{y} = y - y^3, \end{cases}$$

which is easily solved as a singularly perturbed, but separable, equation for y, followed by a linear equation for x. The corresponding inner problem

$$\begin{cases} \dfrac{du}{d\tau} = \epsilon v u, & u(0) = x(0), \\ \dfrac{dv}{d\tau} = v - v^3, & v(0) = y(0), \end{cases}$$

with the stretched time $\tau = t/\epsilon$, has the rapidly varying solution

$$(u_0(\tau), v_0(\tau)) = (x(0), y(0)/\sqrt{y^2(0) - [y^2(0) - 1]e^{-2\tau}})$$

when $\epsilon = 0$ and, therefore, the steady states $(u_0(\infty), v_0(\infty)) = (x(0), 1)$, $(x(0), 0)$, or $(x(0), -1)$ if $y(0) > 0, = 0$, or < 0, respectively. We note that a complete inner expansion

$$\begin{pmatrix} u(\tau, \epsilon) \\ v(\tau, \epsilon) \end{pmatrix} \sim \sum_{j=0}^{\infty} \begin{pmatrix} u_j(\tau) \\ v_j(\tau) \end{pmatrix} \epsilon^j,$$

corresponding to each of these possibilities, could be generated termwise by equating coefficients in the differential equations and the initial conditions in the obvious manner. The coefficients of ϵ^1, for example, require us to take

$$\begin{cases} \dfrac{du_1}{d\tau} = v_0 u_0, & u_1(0) = 0, \\ \dfrac{dv_1}{d\tau} = (1 - 3v_0^2)v_1, & v_1(0) = 0 \,. \end{cases}$$

Thus, $u_1(\tau) = x(0)\int_0^\tau v_0(s)ds$ and $v_1(\tau) = 0$. Writing

$$v_0(\tau) = Y_0(0) + \eta_0(\tau),$$

where $\eta_0 \to 0$ as $\tau \to \infty$,

$$u_1(\tau) = (Y_0(0)\tau + \int_0^\tau \eta_0(s)ds)x(0)$$

will be unbounded as $\tau \to \infty$ unless $Y_0(0) = 0$. Later terms

$$u_k(\tau) = \int_0^\tau [Y_0(0) + \eta_0(s)]u_{k-1}(s)ds,$$

corresponding to $v_k(\tau) = 0$, will grow algebraically at infinity, so we must expect higher-order matching to be complicated.

The corresponding outer expansion

$$\begin{pmatrix} X(t,\epsilon) \\ Y(t,\epsilon) \end{pmatrix} \sim \sum_{j \geq 0} \begin{pmatrix} X_j(t) \\ Y_j(t) \end{pmatrix} \epsilon^j$$

(appropriate for any bounded $t > 0$) will satisfy the system

$$\begin{cases} \dot{X} = YX, \\ \epsilon \dot{Y} = Y - Y^3 \end{cases}$$

on $t > 0$ as a power series in ϵ and it should match the inner solution as $\tau \to \infty$ ($t \to 0^+$). Its zeroth-order term $\begin{pmatrix} X_0 \\ Y_0 \end{pmatrix}$ must satisfy the reduced system and

$$\begin{pmatrix} X_0(0) \\ Y_0(0) \end{pmatrix} = \lim_{\tau \to \infty} \begin{pmatrix} u_0(\tau) \\ v_0(\tau) \end{pmatrix} = \begin{pmatrix} x(0) \\ v_0(\infty) \end{pmatrix}.$$

Hence,

$$Y_0(t) = v_0(\infty)$$

and

$$X_0(t) = e^{Y_0 t} x(0).$$

[If $y(0) > 0$, we must restrict attention to bounded t intervals since X_0 will then be unbounded as $t \to \infty$.] Here, the limiting outer solutions are actually exact solutions of the system, though they will not satisfy the initial condition unless $y^2(0) = 0$ or 1. The $O(\epsilon)$ terms next imply that the first-order coefficients $\begin{pmatrix} X_1 \\ Y_1 \end{pmatrix}$ must satisfy the linear system

$$\begin{cases} \dot{X}_1 = Y_0 X_1 + Y_1 X_0, \\ \dot{Y}_0 = Y_1 - 3Y_0^2 Y_1. \end{cases}$$

Since $\dot{Y}_0 = 0$ and $1 \neq 3Y_0^2$, we will necessarily have

$$Y_1(t) = 0$$

and

$$X_1(t) = e^{Y_0 t} X_1(0).$$

Thus,

$$X(t,\epsilon) = X_0(t) + \epsilon X_1(t) + \cdots = e^{Y_0 t}[x(0) + \epsilon X_1(0) + \cdots].$$

Substituting $t = \epsilon \tau$ and expanding the exponential for finite τ yields

$$X(t,\epsilon) = (1 + \epsilon Y_0 \tau + \cdots)x(0) + \epsilon X_1(0)(1 + \cdots)$$
$$= x(0) + \epsilon[Y_0 x(0)\tau + X_1(0)] + \epsilon^2(\cdots).$$

Recall that the corresponding inner expansion satisfies

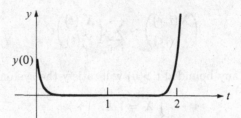

Fig. 2.4. y vs. t for small ϵ.

$$u(\tau,\epsilon) = u_0(\tau) + \epsilon u_1(\tau) + \cdots$$

$$= x(0) + \epsilon \left(Y_0\tau + \int_0^\tau \eta_0(s)ds \right) x(0) + \epsilon^2(\cdots).$$

Thus, the inner and outer expansions will match to order ϵ at $\tau = \infty$ if we select the initial value

$$X_1(0) = \left(\int_0^\infty \eta_0(s)ds \right) x(0).$$

Higher-order matching is also possible, though it is not foolproof to carry out. (We might have also matched the expansions by writing the inner expansion as a function of t.) It is important to realize that behavior away from the layer (i.e., for $t > 0$) depends critically on what happens in the layer [where $t = O(\epsilon)$]. We will later propose a simpler "composite" expansion procedure, which eliminates some of the complications of matching, when it works.

(iii)   The simple linear equation

$$\epsilon \dot{y} = (t-1)y$$

can be easily solved exactly. Since $(t-1) < 0$ for $0 \leq t < 1$, we can anticipate having an $O(\epsilon)$ thick initial layer of nonuniform convergence. The exact solution $y(t,\epsilon) = e^{-(1-t/2)t/\epsilon}y(0)$ is actually symmetric about $t = 1$ and it has $O(\epsilon)$ boundary layers at both $t = 0^+$ and $t = 2^-$. The trivial outer solution holds within $0 < t < 2$, but the solution becomes exponentially large as $\epsilon \to 0$ for $t > 2$, provided $y(0) \neq 0$. Pictorially, we have Figure 2.4. The complicated behavior is due to having a turning point at $t = 1$. That the seemingly dead solution comes to life near $t = 2$ is, indeed, a surprise.

We first note that the initial layer can be simply analyzed by introducing the stretched time

$$\tau = t/\epsilon$$

and seeking a power series expansion of the stretched inner problem

$$\frac{dz}{d\tau} = -(1 - \epsilon\tau)z, \qquad z(0) = y(0)$$

on $\tau \geq 0$. Setting $z(\tau, \epsilon) = z_0(\tau) + \epsilon z_1(\tau) + \cdots$, we will successively require

$$\frac{dz_0}{d\tau} = -z_0, \qquad z_0(0) = y(0),$$

$$\frac{dz_1}{d\tau} = -z_1 + \tau z_0, \quad z_1(0) = 0,$$

etc. Thus,

$$z_0(\tau) = e^{-\tau}y(0),$$

$$z_1(\tau) = \tfrac{1}{2}\tau^2 e^{-\tau}y(0),$$

etc. Note, in particular, that all these terms tend to the trivial outer limit as $\tau \to \infty$.

Beyond the initial layer, we might look for an outer solution

$$Y(t, \epsilon) = Y_0(t) + \epsilon Y_1(t) + \cdots.$$

The differential equation implies that

$$\epsilon(\dot{Y}_0 + \epsilon\dot{Y}_1 + \cdots) = (t - 1)(Y_0 + \epsilon Y_1 + \cdots).$$

Thus, we successively ask that

$$(t - 1)Y_0 = 0,$$

$$(t - 1)Y_1 = \dot{Y}_0,$$

etc. Here, $t = 1$ is a turning point since the reduced equation for Y_0 becomes singular there. Thus, at least on $0 < t < 1$, we will have

$$Y_0(t) = Y_1(t) = \cdots = 0,$$

i.e., a trivial outer expansion. This outer expansion matches the inner expansion at $t = 0$ since $Y(0, \epsilon) = \lim_{\tau \to \infty} z(\tau, \epsilon) = z(\infty, \epsilon)$. That this outer expansion continues to be valid throughout $0 < t < 2$ is lucky, and, perhaps, unexpected. Further formalities show that the exact solution could also be conveniently asymptotically represented in the form

$$y(t, \epsilon) \sim z(\tau, \epsilon) + z(\kappa, \epsilon),$$

where $z(\tau, \epsilon) = e^{-\tau}e^{\epsilon\tau^2/2}y(0)$ provides the initial layer behavior while analogous terminal layer behavior occurs near $t = 2$ as $\kappa = (2 - t)/\epsilon$ tends to infinity. Within $(0, 2)$, of course, both $z(\tau, \epsilon)$ and $z(\kappa, \epsilon)$ are asymptotically negligible. Beyond $t = 2$, both y and $z(\kappa, \epsilon)$ blow up.

(iv)   The *nonlinear* equation

$$\epsilon \dot{y} = (1 - t)y - y^2,$$

with a prescribed positive initial value $y(0)$, was first used by Dahlquist et al. (1982) as a challenging test problem for stiff numerical integrators. Without special intervention or extreme care, most library codes will erroneously provide the limiting solution $Y_0(t) = 1 - t$ beyond $t = 1$ for small values of ϵ. (The function Y_0 does, after all, satisfy the differential equation, except for the small residual $-\epsilon$.)

Fortunately, the initial value problem can be readily integrated as a Riccati equation. The asymptotic behavior follows by expanding the solution

$$y(t, \epsilon) = \left\{ e^{[(1-t)^2 - 1]/2\epsilon} + \frac{1}{\epsilon} y(0) \int_{1-t}^{1} e^{[(1-t)^2 - s^2]/2\epsilon} ds \right\}^{-1} y(0)$$

asymptotically [cf. Erdélyi (1956)] through integration by parts. The reduced equation $(1 - t)Y_0 - Y_0^2 = 0$ has the two solutions $Y_0(t) = 1 - t$ and $Y_0(t) = 0$, which cross at $t = 1$. Unless, $y(0) = 0$ or 1, one must expect an initial layer of nonuniform convergence of $O(\epsilon)$ thickness. Letting $z(\tau, \epsilon)$ satisfy the corresponding inner problem

$$\frac{dz}{d\tau} = (1 - \epsilon\tau)z - z^2, \qquad z(0) = y(0),$$

any limiting inner solution would necessarily satisfy the separable equation $dz_0/d\tau = z_0 - z_0^2$. [If we allowed $y(0) < 0, z_0$ would decrease indefinitely and no limiting solution could be expected.] For $y(0) = 0$, we would get $z_0(\tau) \equiv 0$. For $y(0) > 0$, however, the limiting inner solution $z_0(\tau)$ would tend monotonically to the steady state $z_0(\infty) = 1$ since this rest point is stable. Thus, z_0 will match the limiting outer solution $Y_0(t) = 1 - t$ as $\tau \to \infty$ $(t \to 0^+)$. If we seek a corresponding outer solution

$$Y(t, \epsilon) = Y_0(t) + \epsilon Y_1(t) + \epsilon^2(\cdots)$$

beyond $t = 0$, Y_1 will necessarily satisfy $(1 - t)Y_1 - 2Y_0Y_1 = \dot{Y}_0$, i.e.,

$$(1 - t)Y_1 = 1,$$

and subsequent terms Y_j will satisfy analogous nonhomogeneous linear algebraic equations. All may be singular at the turning point $t = 1$. Until $t = 1$, however, the outer solution is uniquely obtained termwise in the form

$$Y(t, \epsilon) = (1 - t) + \frac{\epsilon}{1 - t} - \frac{2}{(1 - t)^3} \epsilon^2 + \epsilon^3(\cdots).$$

To determine what happens beyond $t = 1$ may be figured out by doing further stretching and matching locally. Since y and Y_0 become small as

$t \to 1$ (while $\dot{y}$ converges nonuniformly from -1 to 0 as t passes through 1), we rescale and look for a local solution of the form

$$y = \epsilon^\alpha w(\lambda),$$

where λ is the stretched variable

$$\lambda = (t-1)/\epsilon^\beta$$

and α and β are unspecified positive parameters. Since $\dot{y} = \epsilon^{\alpha-\beta} dw/d\lambda$ should remain bounded, we pick $\alpha = \beta$. Then, the differential equation implies that $\epsilon \, dw/d\lambda = -\epsilon^{2\alpha} w(w + \lambda)$. We balance the orders of the terms by picking $\alpha = \frac{1}{2}$. Thus, w must satisfy the parameterless equation

$$\frac{dw}{d\lambda} = -w(w + \lambda).$$

Note that it has steady states at $w \sim 0$ and $w + \lambda \sim 0$. Further, $\epsilon^{1/2} w$ must match the outer solution

$$Y(t,\epsilon) = \epsilon^{1/2}\left(-\lambda - \frac{1}{\lambda} + \frac{2}{\lambda^3} + \cdots\right)$$

as $\lambda \to -\infty$, so we must expect the other possible limit $w \sim 0$ as $\lambda \to \infty$. Such a solution w is, indeed, uniquely given by

$$w(\lambda) = e^{-\lambda^2/2}\left(\int_{-\infty}^{\lambda} e^{-s^2/2} ds\right)^{-1}.$$

Thus, $w(\lambda)$ allows a transition (or corner) layer to occur between the outer limit $Y_0(t) = 1 - t$ (valid for $0 < t < 1$) and the trivial limit $Y_0(t) = 0$ (valid for $t > 1$). The outer solution for $t > 1$ can be easily shown to be trivial to all orders ϵ^j, $j > 0$. Our asymptotics has, then, been successful in describing the solution to this initial value problem (for ϵ small, when a numerically obtained "solution" could be misleading). Graphically, we have Figure 2.5.

The idea of using local outer expansions until breakdown and then fixing them up through local expansions is emphasized in the presentation of Carrier (1974). Determining which successive local expansions are appropriate can be quite complicated [cf. Nipp (1988) who used fifteen successive transformations to study the Belousov–Zhabotinskii reaction]. Related problems concerning exchange of stability are examined in the works of Lebovitz and Schaar (1975, 1977).

The matching process used in these examples is valid quite generally. Typically, one defines an outer expansion y_ϵ^o and an inner expansion y_ϵ^i by use of regular perturbation methods and appropriate independent variables t and τ, respectively, which are related by a stretching transformation

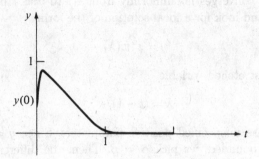

Fig. 2.5. Dahlquist's knee.

such as $\tau = t/\epsilon$. The corresponding outer and inner approximations represent the same solution y in different regions. Presuming, for simplicity, an overlapping domain of validity, one should be able to equate $(y_\epsilon^o)^i$, the outer expansion in the inner region, and $(y_\epsilon^i)^o$, the inner expansion in the outer region, by use of the stretching transformation (for small values of ϵ). Such "matching" [which has been seriously studied by Fraenkel (1969), Eckhaus (1979), and Lagerstrom (1988), among others], when successful, allows us to define a (n additive) composite expansion $y_\epsilon^c = y_\epsilon^o + y_\epsilon^i - (y_\epsilon^i)^o$, which agrees asymptotically with the outer expansion in the outer region and with the inner expansion in the inner region, but which is actually a uniformly valid asymptotic expansion for $y(t, \epsilon)$ [cf. also Van Dyke (1964)]. Observe that matching corresponds roughly to the analytic continuation of functions as used in complex function theory. Matching, in this sense, was often used by nineteenth-century analysts like Schwarz [cf. Chin and Hedstrom (1991)]. Alternatively, note that the difference $y_\epsilon^i - (y_\epsilon^o)^i$ is a boundary layer correction to the outer expansion y_ϵ^o which tends to zero at the edge of the boundary layer. An underlying extension principle due to Kaplun [cf. Kaplun (1967)] seems to be fundamental. For truncated approximations to inner and outer solutions (as we would generate in determining formal solutions to problems), matching is not so simple. The situation, naturally, becomes even more complicated when there are several regions of nonuniform convergence.

D. The Nonlinear Vector Problem (Tikhonov–Levinson Theory)

Now consider the singularly perturbed initial value problem

$$\begin{cases} \dot{x} = f(x, y, t, \epsilon), \\ \epsilon\dot{y} = g(x, y, t, \epsilon) \end{cases}$$

for an m vector x and an n vector y on some bounded interval, say $0 \le t \le 1$, with $x(0)$ and $y(0)$ prescribed. We shall, for simplicity, assume that the functions f and g are infinitely differentiable in their x, y, and t arguments and that they have asymptotic series expansions in ϵ. Such problems were first analyzed by Tikhonov and Levinson and their students in the 1950s [cf. Wasow (1965)]. The corresponding reduced problem consists of the nonlinear differential-algebraic system

$$\begin{cases} \dot{X}_0 = f(X_0, Y_0, t, 0), \\ 0 = g(X_0, Y_0, t, 0) \end{cases}$$

together with the initial condition $X_0(0) = x(0)$. {Such "DAEs" are of independent interest [cf. Campbell (1980, 1982) and Brenan et al. (1989)] and are sometimes solved by introducing a small artificial parameter μ to define a "regularized" problem and then seeking the limiting solution as $\mu \to 0$.} The reduced problem could provide the limiting solution on $0 < t \le 1$ if the corresponding limiting inner problem

$$\frac{dz_0}{d\tau} = g(x(0), z_0, 0, 0), \qquad z_0(0) = y(0),$$

had a bounded solution $z_0(\tau)$ for all $\tau \ge 0$ which matched Y_0 in the sense that

$$z_0(\infty) = Y_0(0).$$

We will say the given problem is *boundary layer stable* if a bounded solution $z_0(\tau)$ to this limiting inner problem exists throughout $\tau \ge 0$ and has a limit at infinity. Occasionally, it happens that limiting solutions do not satisfy the limiting system [cf. the treatment of Lax and Levermore (1983) of weak solutions to the Korteweg–deVries equation].

If g_y is nonsingular along the solution (X_0, Y_0) of the reduced problem, we can differentiate the algebraic constraint $g(X_0, Y_0, t, 0) = 0$ with respect to t to obtain $g_x f + g_y \dot{Y}_0 + g_t = 0$. Thus, Y_0 must satisfy the initial value problem

$$\dot{Y}_0 = -g_y^{-1}(X_0, Y_0, t, 0)[g_x(X_0, Y_0, t, 0)f(X_0, Y_0, t, 0)$$
$$+ g_t(X_0, Y_0, t, 0)], \qquad Y_0(0) = z_0(\infty)$$

which is coupled to the remaining initial value problem

$$\dot{X}_0 = f(X_0, Y_0, t, 0), \qquad X_0(0) = x(0).$$

Note that the initial value problem for X_0 and Y_0 has the same differential order as the original system, but since it is not singularly perturbed, it is not a stiff problem. *Assuming* existence of a solution $\binom{X_0}{Y_0}$ throughout the interval of interest, $0 \le t \le 1$, approximating it numerically should then be easy. (We note, however, that Gu (1991) has initiated the study of boundary layer systems with only finite intervals of existence.) Sometimes, we can directly find a solution

$$Y_0 = \phi(X_0, t)$$

of the algebraic system $g(X_0, Y_0, t, 0) = 0$, starting at the point $(x(0), z_0(\infty), 0)$. Then our limiting solution is obtained by simply solving the reduced-order initial value problem

$$\dot{X}_0 = f(X_0, \phi(X_0, t), t, 0) \equiv \mathcal{F}(X_0, t), \quad X_0(0) = x(0)$$

to obtain $X_0(t)$ and thereby $Y_0(t)$. The implicit function theorem, indeed, guarantees the existence of a locally unique root ϕ, as long as $g_y(X_0(t), Y_0(t), t, 0)$ remains nonsingular, but it does not provide any simple way of solving $g = 0$ for Y_0. Note that $g = 0$ may have other solutions, but the only right one to use is the isolated root selected through the limiting boundary layer stability problem, i.e., the solution $\binom{X_0}{Y_0}$ passing through $(x(0), Y_0(0), 0)$ where the initial value $Y_0(0) = z_0(\infty)$ is obtained by integrating the initial value problem for $z_0(\tau)$ from $\tau = 0$ to ∞. If g_y becomes singular during the t integration, our procedure for obtaining the outer limit generally breaks down. Such points are analogous to the turning points already encountered for linear examples.

Successful integration of the limiting inner problem for $z_0(\tau)$ relies on stability considerations. We will *assume* that

$$(z - z_0(\infty))'g(x(0), z, 0, 0) \leq -\kappa(z - z_0(\infty))'(z - z_0(\infty))$$

holds for all relevant z and for some $\kappa > 0$. We note that κ plays the role of a one-sided Lipschitz constant, and that such "dissipation" conditions are common in the stiff equations literature [cf. Dahlquist (1959)] and elsewhere. We will show that

$$z_0(\tau) = z_0(\infty) + O(e^{-\kappa\tau})$$

will then hold for all $\tau \geq 0$ since $\eta_0(\tau) \equiv z_0(\tau) - z_0(\infty)$ will satisfy the initial value problem $d\eta_0/d\tau = g(x(0), \eta_0 + z_0(\infty), 0, 0), \eta_0(0) = y(0) - z_0(\infty)$. Using the inner product norm $\|\eta_0\| = \sqrt{\eta_0'\eta_0}$, we will have

$$\frac{d}{d\tau}\|\eta_0\|^2 = 2\frac{d\eta_0'}{d\tau}\eta_0 = 2\eta_0'g(x(0), \eta_0 + z_0(\infty), 0, 0) \leq -2\kappa\|\eta_0\|^2,$$

so, upon integration, the existence of a solution $\eta_0(\tau)$ on $\tau \geq 0$ is guaranteed which satisfies

$$\|\eta_0(\tau)\| \leq e^{-\kappa\tau}\|\eta_0(0)\| \qquad \text{for all } \tau \geq 0.$$

Because

$$\eta' g(x(0), \eta + z_0(\infty), 0, 0) = \eta' \left(\int_0^1 g_y(x(0), s\eta + z_0(\infty), 0, 0) ds \right) \eta,$$

our stability hypothesis would hold under the *classical assumption* that $g_y(x(0), z, 0, 0)$ remains strictly stable for all z. To simply treat the linearized problem, let us instead *assume* that

$$g_y(x(0), z_0(\tau), 0, 0)$$

remains strictly stable for all $\tau \geq 0$. This will, in particular, imply the stability of $g_y(x(0), z_0(\infty), 0, 0) = g_y(X_0(0), Y_0(0), 0, 0)$.

In an analogous fashion, it will be convenient to guarantee that the limiting outer solution $\binom{X_0(t)}{Y_0(t)}$ remains attractive to small perturbations throughout $0 \leq t \leq 1$. We will guarantee stability of this limiting outer solution by *assuming* that

$$g_y(X_0(t), Y_0(t), t, 0)$$

remains a strictly stable matrix for $0 \leq t \leq 1$.

Under these hypotheses, we shall obtain an asymptotic solution to our initial value problem in the form

$$\begin{cases} x(t, \epsilon) = X(t, \epsilon) + \epsilon\xi(\tau, \epsilon), \\ y(t, \epsilon) = Y(t, \epsilon) + \eta(\tau, \epsilon) \end{cases}$$

with an outer expansion

$$\binom{X(t, \epsilon)}{Y(t, \epsilon)} \sim \sum_{j=0}^{\infty} \binom{X_j(t)}{Y_j(t)} \epsilon^j$$

and an initial layer correction $\binom{\epsilon\xi}{\eta}$ satisfying

$$\binom{\xi(\tau, \epsilon)}{\eta(\tau, \epsilon)} \sim \sum_{j=0}^{\infty} \binom{\xi_j(\tau)}{\eta_j(\tau)} \epsilon^j$$

whose terms (ξ_j, η_j) all decay to zero as $\tau = t/\epsilon$ tends to infinity. Thus, the outer expansion must be a smooth solution of the system

$$\begin{cases} \dot{X} = f(X, Y, t, \epsilon), \\ \epsilon\dot{Y} = g(X, Y, t, \epsilon). \end{cases}$$

Equating coefficients successively, we first find that (X_0, Y_0) must satisfy the nonlinear reduced system

$$\begin{cases} \dot{X}_0 = \mathcal{F}(X_0, t), \quad X_0(0) = x(0), \\ Y_0 = \phi(X_0, t) \end{cases}$$

for the root ϕ of $g(X_0, \phi(X_0, t), t, 0) = 0$ obtained by matching with the limiting inner solution at $t = 0^+$. Later terms (X_j, Y_j) must likewise satisfy linear differential-algebraic systems of the form

$$\begin{cases} \dot{X}_j = f_x(X_0, Y_0, t, 0)X_j + f_y(X_0, Y_0, t, 0)Y_j + \tilde{f}_{j-1}(t), \\ 0 = g_x(X_0, Y_0, t, 0)X_j + g_y(X_0, Y_0, t, 0)Y_j + \tilde{g}_{j-1}(t), \end{cases}$$

where $\tilde{f}_{j-1}$ and $\tilde{g}_{j-1}$ are known recursively in terms of t and the preceding coefficients X_k and Y_k for $k < j$ and their derivatives. Since differentiation of $g(X_0, \phi, t, 0) = 0$ with respect to X_0 implies that $\phi_x = -g_y^{-1} g_x$, we can solve the latter algebraic system to obtain

$$Y_j = \phi_x(X_0, t)X_j + \hat{g}_{j-1}(t),$$

leaving the linear system

$$\dot{X}_j = \mathcal{F}_x(X_0, t)X_j + \hat{f}_{j-1}(t)$$

for X_j since differentiation of $\mathcal{F}(X, t) = f(X, \phi, t, 0)$ implies that $\mathcal{F}_x = f_x + f_y \phi_x$. Note that the homogeneous system $dp/dt = \mathcal{F}_x(X_0, t)p$ is the variational equation for X_0. It follows that the outer solution $\binom{X(t,\epsilon)}{Y(t,\epsilon)}$ can be completely determined asymptotically once the initial value $X(0, \epsilon)$ is specified. The initial condition for X_j will be determined by the previous term ξ_{j-1} in the initial layer correction since the ϵ^j coefficient in the initial condition $x(0) = X(0, \epsilon) + \epsilon\xi(0, \epsilon)$ implies that

$$X_j(0) = -\xi_{j-1}(0) \qquad \text{for each } j > 0.$$

The ϵ^j coefficient in the other initial condition $y(0) = Y(0, \epsilon) + \eta(0, \epsilon)$ then specifies

$$\eta_j(0) = -Y_j(0)$$

which is already completely specified by $X_j(0)$. The initial value $\eta(0, \epsilon)$, as we will next learn, will be adequate for the termwise determination of the initial layer correction.

Differentiating the assumed form of the asymptotic solution implies that $\dot{x} = \dot{X} + d\xi/d\tau = f$ and $\epsilon\dot{y} = \epsilon\dot{Y} + d\eta/d\tau = g$. Because (x, y) and (X, Y) both satisfy the original system, it follows that the initial layer correction $(\epsilon\xi, \eta)$ must be obtained as a decaying solution of the nonlinear system

$$\begin{cases} \dfrac{d\xi}{d\tau} = f(X + \epsilon\xi, Y + \eta, \epsilon\tau, \epsilon) - f(X, Y, \epsilon\tau, \epsilon), \\[2mm] \dfrac{d\eta}{d\tau} = g(X + \epsilon\xi, Y + \eta, \epsilon\tau, \epsilon) - g(X, Y, \epsilon\tau, \epsilon) \end{cases}$$

on the semi-infinite interval $\tau \geq 0$. In particular, then, the leading terms will have to satisfy the nonlinear system

$$\begin{cases} \dfrac{d\xi_0}{d\tau} = f(x(0), Y_0(0) + \eta_0, 0, 0) - f(x(0), Y_0(0), 0, 0), \\[2mm] \dfrac{d\eta_0}{d\tau} = g(x(0), Y_0(0) + \eta_0, 0, 0) - g(x(0), Y_0(0), 0, 0) \end{cases}$$

on $\tau \geq 0$ and decay to zero as $\tau \to \infty$. Later terms must satisfy a linearized system

$$\begin{cases} \dfrac{d\xi_k}{d\tau} = f_y(x(0), Y_0(0) + \eta_0(\tau), 0, 0)\eta_k + p_{k-1}(\tau), \\[2mm] \dfrac{d\eta_k}{d\tau} = g_y(x(0), Y_0(0) + \eta_0(\tau), 0, 0)\eta_k + q_{k-1}(\tau), \end{cases}$$

where the p_{k-1}'s and q_{k-1}'s are known successively. We will show that these terms all decay to zero exponentially (roughly like $e^{-\kappa\tau}$) as $\tau \to \infty$.

Note that $z_0(\tau) = \eta_0(\tau) + Y_0(0)$ will satisfy the limiting inner problem

$$\frac{dz_0}{d\tau} = g(x(0), z_0, 0, 0), \qquad z_0(0) = y(0).$$

Under our boundary layer stability hypothesis, $z_0(\tau)$ will have a unique solution for $\tau \geq 0$ which decays exponentially to $Y_0(0)$ as $\tau \to \infty$. Knowing $z_0(\tau)$, we obtain $\eta_0(\tau) = z_0(\tau) - Y_0(0)$ and thereby $d\xi_0/d\tau$. Because $\xi_0(\infty) = 0$, we must then take

$$\xi_0(\tau) = -\int_{\tau}^{\infty} [f(x(0), z_0(s), 0, 0) - f(x(0), z_0(\infty), 0, 0)]ds.$$

Note that ξ_0, η_0, and their derivatives will all decay exponentially to zero as $\tau \to \infty$. Moreover, ξ_0 determines $X_1(0) = -\xi_0(0)$ and, thereby, $X_1(t)$ and $Y_1(t)$. It also enables us to specify a linear initial value problem

$$\frac{d\eta_1}{d\tau} = g_y(x(0), z_0(\tau), 0, 0)\eta_1 + q_0(\tau), \qquad \eta_1(0) = -Y_1(0)$$

for $\eta_1(\tau)$. Using variation of parameters, η_1 can be obtained in terms of the fundamental matrix $Q(\tau)$ for the linearized (or variational) system

$$\frac{dQ}{d\tau} = g_y(x(0), z_0(\tau), 0, 0)Q, \qquad Q(0) = I,$$

i.e.,

$$\eta_1(\tau) = -Q(\tau)Y_1(0) + \int_{0}^{\tau} Q(\tau)Q^{-1}(s)q_0(s)ds.$$

We observe that our boundary layer stability hypothesis implies that

$$Q(\tau) = O(e^{-\kappa\tau})$$

as $\tau \to \infty$, as could be verified by using a successive approximations argument for the corresponding integral equation. The exponential decay of Q implies that of η_1. It also yields

$$\xi_1(\tau) = -\int_\tau^\infty [f_y(x(0), z_o(s), 0, 0)\eta_1(s) + p_0(s)]ds,$$

so we have specified the initial vector $X_2(0) = -\xi_1(0)$ needed to obtain the second-order terms in the outer expansion. Later terms follow successively in the same manner. We note that Hoppensteadt (1966) showed that such results continue to hold throughout $t \geq 0$, provided $\binom{X_0}{Y_0}$ is asymptotically stable as $\tau \to \infty$. Smith (1985) calls our procedure the O'Malley/Hoppensteadt construction.

In closing, we list some direct applications.

1. The expansion procedure justifies the very successful and much used Michaelis–Menton theory (on $t > 0$) and indicates higher-order improvements as well.

2. For the linear system

$$\begin{cases} \dot{x} = A(t)x + B(t)y, \\ \epsilon\dot{y} = C(t)x + D(t)y, \end{cases}$$

of Section 2B, the reduced system

$$\begin{cases} \dot{X}_0 = A(t)X_0 + B(t)Y_0, \quad X_0(0) = x(0), \\ 0 = C(t)X_0 + D(t)Y_0 \end{cases}$$

has a unique solution as long as $D(t)$ remains a stable matrix. The corresponding limiting inner system

$$\frac{dz_0}{d\tau} = C(0)x(0) + D(0)z_0, \qquad z_0(0) = y(0)$$

has the unique exponentially decaying solution

$$z_0(\tau) = e^{D(0)\tau}[y(0) + D^{-1}(0)C(0)x(0)] - D^{-1}(0)C(0)x(0)$$

since $D(0)$ is stable. Note that the steady state $z_0(\infty)$ matches the outer limit $Y_0(0)$. Higher-order matching is carried out in a similar fashion.

3. For the system

$$\begin{cases} \dfrac{dx}{dt} = yx, \\ \epsilon\dfrac{dy}{dt} = y - y^3 \end{cases}$$

of two nonlinear scalar equations, the reduced problem

$$\begin{cases} \dfrac{dX_0}{dt} = Y_0 X_0, \qquad X_0(0) = x(0), \\ 0 = Y_0 - Y_0^3 \end{cases}$$

will apply, provided boundary layer stability and stability of the reduced solution both hold. These conditions would be guaranteed if
$g_y = 1 - 3y^2$ remains negative in the domain of interest, as well as by
somewhat weaker hypotheses.

For $y(0) > 0$, the boundary layer problem

$$\frac{dz_0}{d\tau} = z_0 - z_0^3, \qquad z_0(0) = y(0)$$

has a unique exponentially decaying solution on $\tau \geq 0$ tending to
$z_0(\infty) = Y_0(0) = 1$ as $\tau \to \infty$. The corresponding outer limit $Y_0(t) \equiv 1$
will be stable on $0 \leq t \leq 1$ since $g_y = 1 - 3Y_0^2 = -2$ there. Thus, the
limiting solution for $t > 0$ is given by $\binom{X_0}{Y_0} = \binom{e^t x(0)}{1}$. Likewise, if
$y(0) < 0$, the limit is $\binom{X_0}{Y_0} = \binom{e^{-t} x(0)}{-1}$ for $t > 0$, along which we will
also have $g_y = -2$.

4.   In similar fashion, our theory implies that the initial value problem for
$\epsilon \dot{y} = (t - 1)y$ has the trivial limit on $0 < t < 1$ since $g_y = t - 1 < 0$
there. Likewise, the limiting solution $Y_0(t) = 1 - t$ corresponding to the
initial value problem for $\epsilon \dot{y} = (1 - t)y - y^2$ holds on $0 < t < 1$ provided
the limiting boundary layer problem $dz_0/d\tau = z_0 - z_0^2, z_0(0) = y(0)$,
has the steady state $z_0(\infty) = Y_0(0) = 1$ [i.e., provided $y(0) > 0$], since
$g_y = (1 - t) - 2Y_0 = -(1 - t) < 0$ there.

The assumptions we've used can be modified somewhat, but not
greatly, as the following examples illustrate.

a.   For the linear problem

$$\epsilon \dot{y} = -t(y + a(t)),$$

the initial value problem can be easily solved using the integrating
factor $e^{t^2/2\epsilon}$. In the initial layer, the appropriate stretched variable is
now

$$\kappa = t/\sqrt{\epsilon}$$

and the corresponding limiting problem is

$$\frac{dz_0}{d\kappa} = -\kappa(z_0 + a(0)).$$

Its solution $z_0(\kappa) = -a(0) + e^{-\kappa^2/2}(y(0) + a(0))$ has a steady state
$z_0(\infty)$ which matches the limiting outer solution $Y_0(t) = -a(t)$ as

$t \to 0^+$. Note that the initial point $t = 0$ is a turning point, where $Y_0(t)$ loses stability because $g_y = -t$ becomes zero. As a result, the initial layer is thicker than the $O(\epsilon)$ layers encountered when there was no turning point. The solution, indeed, has the asymptotic form

$$y(t, \epsilon) = Y(t, \sqrt{\epsilon}) + \zeta(\kappa, \sqrt{\epsilon}),$$

where $Y(t, \sqrt{\epsilon}) \sim Y_0(t)$ as $\epsilon \to 0$ and $\zeta \to 0$ as $\kappa = t/\sqrt{\epsilon} \to \infty$. Moreover, the outer solution Y and the initial layer correction ζ both have expansions in powers of $\sqrt{\epsilon}$.

b. For linear systems

$$\epsilon \dot{y} = A(\epsilon)y,$$

the preceding theory applies if $A(0)$ is stable. Now consider the example with

$$A(\epsilon) = \begin{pmatrix} 1 - 2\epsilon & 2 - 2\epsilon \\ -1 + \epsilon & -2 + \epsilon \end{pmatrix} \equiv A_0 + A_1\epsilon.$$

Since $A(\epsilon)$ has eigenvalues -1 and $-\epsilon$ and corresponding eigenvectors $\begin{pmatrix} 1 \\ -1 \end{pmatrix}$ and $\begin{pmatrix} 2 \\ -1 \end{pmatrix}$, the general solution is a linear combination of $\begin{pmatrix} 1 \\ -1 \end{pmatrix}e^{-t/\epsilon}$ and $\begin{pmatrix} 2 \\ -1 \end{pmatrix}e^{-t}$. Alternatively, the initial value problem has the explicit solution

$$y(t, \epsilon) = e^{A(\epsilon)t/\epsilon}y(0) = \begin{pmatrix} e^{-t/\epsilon} & 2e^{-t} \\ -e^{-t/\epsilon} & -e^{-t} \end{pmatrix} \begin{pmatrix} 1 & 2 \\ -1 & -1 \end{pmatrix}^{-1} y(0)$$

which can, more conveniently, be expressed in the form

$$y(t, \epsilon) = Y_0(t) + \eta_0(\tau)$$

where Y_0 satisfies the reduced problem

$$A_0 Y_0 = 0$$

and $\eta_0 \to 0$ as $\tau = t/\epsilon \to \infty$. Because A_0 has rank one, Y_0 must be of the form

$$Y_0(t) = \begin{pmatrix} 2 \\ -1 \end{pmatrix}\alpha_0(t)$$

for an unspecified α_0. The limiting inner problem $dz_0/d\tau = A_0z_0$, $z_0(0) = y(0)$, likewise has the unique solution

$$z_0(\tau) = \begin{pmatrix} e^{-\tau} & 2 \\ -e^{-\tau} & -1 \end{pmatrix} \begin{pmatrix} -1 & -2 \\ 1 & 1 \end{pmatrix} y(0),$$

so its steady state matches $Y_0(0)$ if $\alpha_0(0) = y_1(0) + y_2(0)$. The $O(\epsilon)$ terms in the outer equation $\epsilon \dot{Y} = (A_0 + \epsilon A_1)Y$ require that

$$A_0 Y_1 + A_1 Y_0 = \dot{Y}_0,$$

so solvability requires $\dot{Y}_0 - A_1 Y_0 = \binom{2}{-1}(\dot{\alpha}_0 + \alpha_0)$ to lie in the range of A_0. Thus, we need $2(\dot{\alpha}_0 + \alpha_0) = \dot{\alpha}_0 + \alpha_0$, so $\alpha_0(t) = e^{-t}\alpha_0(0)$ and we have finally determined the unique limiting solution for $t > 0$ to be

$$Y_0(t) = \binom{2}{-1} e^{-t}(y_1(0) + y_2(0)).$$

Such "singular" singular perturbation problems will be discussed in greater generality in Section 2J below.

c. For the equation

$$\epsilon \dot{y} = 1,$$

the Jacobian g_y is not stable and the reduced equation $0 = 1$ is even inconsistent. The initial value problem has the unbounded solution

$$y(t, \epsilon) = \frac{t}{\epsilon} + y(0)$$

for $t > 0$. We note that the rescaled problem for $z = \epsilon y$ is more straightforward to deal with than the original.

d. Consider the linear system

$$\epsilon \dot{y} = D(t, \epsilon)y \equiv U^{-1}(t, \epsilon)AU(t, \epsilon)y,$$

where A is the stable matrix $\begin{pmatrix} -1 & 5 \\ 0 & -1 \end{pmatrix}$ and U is the rapidly varying orthogonal matrix

$$U(t, \epsilon) = \begin{pmatrix} \cos t/\epsilon & -\sin t/\epsilon \\ \sin t/\epsilon & \cos t/\epsilon \end{pmatrix}.$$

Thus, the Jacobian matrix $g_y(x, y, t, \epsilon) = D(t, \epsilon)$ is always bounded, having -1 as a double eigenvalue. $D(t, \epsilon)$ is, however, not a smooth function of t and ϵ. It is, rather, a smooth function of their quotient, and $\dot{D}$ is unbounded like $1/\epsilon$. The nonsingular transformation

$$w = U(t, \epsilon)y$$

converts our problem to a constant coefficient system

$$\epsilon \dot{w} = Bw$$

since $\epsilon \dot{w} = (\epsilon \dot{U} + UD)U^{-1}w$ yields

$$B = A + \epsilon \dot{U}U^{-1} = \begin{pmatrix} -1 & 4 \\ 1 & -1 \end{pmatrix}$$

with the eigenvalues -3 and 1. The occurrence of the unstable eigenvalue 1 implies that some solutions y grow like $e^{t/\epsilon}$. Thus, the eigenvalue stability of g_y is not sufficient to guarantee boundary layer stability when the state matrix D is not smooth. We note that related examples were presented in Vinograd (1952), Hoppensteadt (1966), Coppel (1978), and Kreiss (1978).

E. An Outline of a Proof of Asymptotic Correctness

To show that the solution we have formally generated is asymptotically correct, we take some integer $N > 0$ and set

$$\begin{cases} x(t,\epsilon) = X^N(t,\epsilon) + \epsilon^{N+1} R(t,\epsilon), \\ y(t,\epsilon) = Y^N(t,\epsilon) + \epsilon^{N+1} S(t,\epsilon), \end{cases}$$

where

$$X^N(t,\epsilon) = X_0(t) + \sum_{j=1}^{N} [X_j(t) + \xi_{j-1}(t/\epsilon)]\epsilon^j$$

and

$$Y^N(t,\epsilon) = \sum_{j=0}^{N} [Y_j(t) + \eta_j(t/\epsilon)]\epsilon^j,$$

and prove that the scaled remainders $R(t,\epsilon)$ and $S(t,\epsilon)$ are unique and bounded vectors throughout $0 \le t \le 1$. This is a restatement of the definition of a generalized asymptotic expansion [cf. Erdélyi (1956)]. We note that the specifications of (i) the outer expansion terms (X_j, Y_j) and of (ii) the initial layer correction terms (ξ_j, η_j) imply that (X^N, Y^N) is an ϵ^N approximate solution to our initial value problem in the sense that

$$\begin{cases} \dfrac{dX^N}{dt} = f(X^N, Y^N, t, \epsilon) + \epsilon^N p_1(t,\epsilon), \quad X^N(0,\epsilon) = x(0) + \epsilon^{N+1} q_1(\epsilon), \\ \epsilon\dfrac{dY^N}{dt} = g(X^N, Y^N, t, \epsilon) + \epsilon^N p_2(t,\epsilon), \quad Y^N(0,\epsilon) = y(0) + \epsilon^{N+1} q_2(\epsilon) \end{cases}$$

for bounded functions p_i and q_i. This follows since the first $N + 1$ terms in the outer expansion satisfy the nonlinear system in t with an $O(\epsilon^{N+1})$ residual, while (X^N, Y^N) satisfies the corresponding system in τ with an $O(\epsilon^{N+1})$ error. Since the boundary layer correction terms decay asymptotically, we'll actually have $p_i(t,\epsilon) = O(\epsilon) + O(e^{-\kappa t/\epsilon})$, so the p_i's will have $O(\epsilon)$ t-integrals. Differentiating the representations for x and y, it follows that

$$\begin{cases} \dfrac{dR}{dt} = \dfrac{1}{\epsilon^{N+1}} \left[f(X^N + \epsilon^{N+1} R, Y^N + \epsilon^{N+1} S, t, \epsilon) - f(X^N, Y^N, t, \epsilon) \right] \\ \qquad\qquad - \dfrac{1}{\epsilon} p_1(t,\epsilon), \quad R(0,\epsilon) = -q_1(\epsilon) \\ \epsilon\dfrac{dS}{dt} = \dfrac{1}{\epsilon^{N+1}} \left[g(X^N + \epsilon^{N+1} R, Y^N + \epsilon^{N+1} S, t, \epsilon) - g(X^N, Y^N, t, \epsilon) \right] \\ \qquad\qquad - \dfrac{1}{\epsilon} p_2(t,\epsilon), \quad S(0,\epsilon) = -q_2(\epsilon). \end{cases}$$

Now, we will expand f and g about (X^N, Y^N, t, ϵ) to, for example, obtain

$$f(x, y, t, \epsilon) = f(X^N, Y^N, t, \epsilon) + \epsilon^{N+1} A(t, \epsilon) R + \epsilon^{N+1} B(t, \epsilon) S$$
$$+ \epsilon^{2(N+1)} E(R, S, t, \epsilon),$$

where $A(t, \epsilon) = f_x(X^N, Y^N, t, \epsilon)$, $B(t, \epsilon) = f_y(X^N, Y^N, t, \epsilon)$, and $E(R, S, t, \epsilon)$ is the scaled remainder in the Taylor series. Thus, the vectors R and S will satisfy the nearly linear initial value problem

$$\begin{cases} \dfrac{dR}{dt} = A(t, \epsilon) R + B(t, \epsilon) S - \dfrac{1}{\epsilon} p_1(t, \epsilon) + \epsilon^{N+1} E(R, S, t, \epsilon), \\[2mm] \quad R(0, \epsilon) = -q_1(\epsilon), \\[2mm] \epsilon \dfrac{dS}{dt} = C(t, \epsilon) R + D(t, \epsilon) S - \dfrac{1}{\epsilon} p_2(t, \epsilon) + \epsilon^{N+1} F(R, S, t, \epsilon), \\[2mm] \quad S(0, \epsilon) = -q_2(\epsilon). \end{cases}$$

We note that here the coefficients are not smooth functions of t, but are instead the sum of a smooth function of t and of an initial layer function of $\tau = t/\epsilon$ which decays exponentially to zero as $\tau \to \infty$. The more critical fact is that our stability assumptions imply that the eigenvalues of $D(t, \epsilon) = g_y(X^N, Y^N, t, \epsilon)$ have real parts which are less than $-\kappa < 0$ throughout $0 \leq t \leq 1$ for ϵ sufficiently small. From these facts it follows that the homogeneous linear system

$$\frac{d}{dt} \begin{pmatrix} z_1 \\ \epsilon z_2 \end{pmatrix} = \begin{pmatrix} A & B \\ C & D \end{pmatrix} \begin{pmatrix} z_1 \\ z_2 \end{pmatrix}$$

has an m-dimensional manifold of smooth bounded outer solutions which we will take to be the columns of an $(m + n) \times m$ matrix $\begin{pmatrix} z_{11} \\ z_{21} \end{pmatrix}$ and an n-dimensional manifold of initial layer solutions which we will take to be the columns of an $(m + n) \times n$ matrix $\begin{pmatrix} \epsilon z_{12} \\ z_{22} \end{pmatrix}$ whose entries are all $O(e^{-\kappa t/\epsilon})$ as $t/\epsilon \to \infty$. (For more details, recall our previous discussion of linear systems.) Thus, we have a fundamental matrix

$$z(t, \epsilon) = \begin{pmatrix} z_{11} & \epsilon z_{12} \\ z_{21} & z_{22} \end{pmatrix}$$

for this homogeneous linear system and the corresponding Green's function matrix

$$K(t, s, \epsilon) = z(t, \epsilon) z^{-1}(s, \epsilon) = \begin{pmatrix} K_{11} & \epsilon K_{12} \\ K_{21} & K_{22} \end{pmatrix}$$

for the nonhomogenous initial value problem. Here $K_{11}, K_{12},$ and K_{21} are bounded and $K_{22} = O(\epsilon) + O(e^{-\kappa(t-s)/\epsilon})$ for $t \geq s$. Thus, variation of parameters implies that R and S must satisfy the integral equation

$$\begin{pmatrix} R(t, \epsilon) \\ S(t, \epsilon) \end{pmatrix} = -K(t, 0, \epsilon) \begin{pmatrix} q_1(\epsilon) \\ q_2(\epsilon) \end{pmatrix} - \frac{1}{\epsilon^2} \int_0^t K(t, u, \epsilon) \begin{pmatrix} \epsilon p_1(u, \epsilon) \\ p_2(u, \epsilon) \end{pmatrix} du$$

$$+ \epsilon^N \int_0^t K(t, u, \epsilon) \begin{bmatrix} \epsilon E(R(u, \epsilon), S(u, \epsilon), u, \epsilon) \\ F(R(u, \epsilon), S(u, \epsilon), u, \epsilon) \end{bmatrix} du.$$

It is critical to recall that each $(1/\epsilon)p_i(u,\epsilon)$ and $(1/\epsilon)K_{22}(t,u,\epsilon)$ has a bounded integral over $0 \leq u \leq t \leq 1$. A straightforward fixed point argument shows that the integral equation has a unique bounded solution (R,S) on $0 \leq t \leq 1$ provided ϵ is sufficiently small. This proves that the formal asymptotic expansion we obtained approximates an exact solution of the differential system asymptotically as $\epsilon \to 0$.

Exercise

Consider the initial value problem for the scalar equation

$$\epsilon \dot{y} = a(t,\epsilon)y + b(t,\epsilon)$$

on the interval $0 \leq t \leq 2$ where $a(t,\epsilon) < 0$. Suppose that a and b are arbitrarily smooth on $0 \leq t \leq 1$ and on $1 < t \leq 2$, but that they jump at $t = 1^+$. Thus, we write

$$\begin{pmatrix} a(t,\epsilon) \\ b(t,\epsilon) \end{pmatrix} = \begin{cases} \begin{pmatrix} \alpha_1(t,\epsilon) \\ \beta_1(t,\epsilon) \end{pmatrix} & \text{for } 0 \leq t \leq 1, \\[2mm] \begin{pmatrix} \alpha_2(t,\epsilon) \\ \beta_2(t,\epsilon) \end{pmatrix} & \text{for } 1 < t \leq 2. \end{cases}$$

Show that the asymptotic solution has the form

$$y(t,\epsilon) = \begin{cases} Y_1(t,\epsilon) + \xi(\tau,\epsilon) & \text{for } 0 \leq t \leq 1, \\ Y_2(t,\epsilon) + \eta(\sigma,\epsilon) & \text{for } 1 \leq t \leq 2, \end{cases}$$

where Y_1 and Y_2 are smooth and where $\xi \to 0$ as $\tau = t/\epsilon \to \infty$ and $\eta \to 0$ as $\sigma = (t-1)/\epsilon \to \infty$. In particular, determine the first few terms in the expansions and check the smoothness of y at $t = 1$. Also consider the situation when

$$\begin{pmatrix} a(t,\epsilon) \\ b(t,\epsilon) \end{pmatrix} = \begin{cases} \begin{pmatrix} \alpha_1(t,\epsilon) \\ \beta_1(t,\epsilon) \end{pmatrix}, & 0 \leq t \leq 1, \\[2mm] \begin{pmatrix} \alpha_2(t,\epsilon) + \gamma(\sigma,\epsilon) \\ \beta_2(t,\epsilon) + \delta(\sigma,\epsilon) \end{pmatrix}, & 1 < t \leq 2, \end{cases}$$

where the α_i and β_i are smooth functions of t, and γ and $\delta \to 0$ as $\sigma = (t-1)/\epsilon \to \infty$.

F. Numerical Methods for Stiff Equations

Consider the two-dimensional linear system

$$y' = Ay$$

where the constant matrix

$$A = \frac{1}{2} \begin{pmatrix} \lambda_1 + \lambda_2 & \lambda_1 - \lambda_2 \\ \lambda_1 - \lambda_2 & \lambda_1 + \lambda_2 \end{pmatrix}$$

has eigenvalues λ_1 and λ_2 with $\lambda_2 \ll \lambda_1 < 0$. (To obtain the usual singular perturbation formulation, we could introduce the small parameter $\epsilon = -1/\lambda_2$.) Using the decomposition $A = MDM^{-1}$ with $D = \begin{pmatrix} \lambda_1 & 0 \\ 0 & \lambda_2 \end{pmatrix}$ and $M = \begin{pmatrix} 1 & 1 \\ 1 & -1 \end{pmatrix}$, the general solution of the system is given by $y(t) = Me^{Dt}k$ for any constant vector k, or as

$$y(t) = \begin{pmatrix} 1 \\ 1 \end{pmatrix} e^{\lambda_1 t} k_1 + \begin{pmatrix} 1 \\ -1 \end{pmatrix} e^{\lambda_2 t} k_2$$

for arbitrary scalars k_1 and k_2. If we tried to integrate the initial value problem using Euler's method with a constant stepsize $h > 0$, we would use the difference quotient

$$\frac{y_{n+1} - y_n}{h} = Ay_n$$

so

$$y_{n+1} = (I + hA)y_n \qquad \text{for every } n.$$

Solving these difference equations, we obtain

$$y_n = \begin{pmatrix} 1 \\ 1 \end{pmatrix} (1 + h\lambda_1)^n c_1 + \begin{pmatrix} 1 \\ -1 \end{pmatrix} (1 + h\lambda_2)^n c_2$$

for constants c_1 and c_2. Stability as $n \to \infty$ forces us to restrict the stepsize h so that

$$|1 + h\lambda_i| \leq 1 \qquad \text{for } i = 1 \text{ and } 2.$$

If $\lambda_1 = -1$ and $\lambda_2 = -10^6$, this requires h to satisfy

$$0 \leq h \leq 2 \times 10^{-6}.$$

By contrast, the truncation error will be bounded by

$$\tfrac{1}{2} h^2 \max_t |y''(t)|,$$

so a tolerable accuracy will be guaranteed on, say, $0 \leq t \leq 10^{-5}$ if we take a small enough stepsize h, even if $y'' = O(10^{12})$ in this region of rapid change. For $10^{-5} \leq t \leq 1$, however, y'' is bounded, so one would hope that the stepsize h could then be made moderately large to achieve the same accuracy. One calls the initial value problem stiff in the *latter* region where stability (rather than accuracy) restricts the stepsize. There, the transient part of the solution $\begin{pmatrix} 1 \\ -1 \end{pmatrix} e^{\lambda_2 t} k_2$ is negligible and the solution behaves like its slow part $\begin{pmatrix} 1 \\ 1 \end{pmatrix} e^{\lambda_1 t} k_1$. To paraphrase Dekker and Verver (1984), the essence of *stiffness* is that the solution to be computed is slowly varying, whereas any perturbations which occur are rapidly damped.

More generally, if we attempt to solve the nonlinear initial value problem

$$y' = f(y, t), \qquad y(0) = y_0,$$

we might find it natural to relate the behavior of the solution to the time constants $1/|\lambda_i|$ or $1/\text{Re}\,\lambda_i$, where λ_i is an eigenvalue of f_y evaluated along the solution. Indeed, Dahlquist (1959) showed that the errors and differences between any two solutions y and $\tilde{y}$ will satisfy

$$\|\tilde{y}(t_2) - y(t_2)\| \leq e^{\int_{t_1}^{t_2} \nu(s)ds} \|\tilde{y}(t_1) - y(t_1)\|$$

where $\nu(t)$ is an upper bound for the logarithmic norm μ of f_y [cf. Coppel (1965)], i.e., $\mu[f_y] = \max_{\mathcal{J} \neq 0}(\mathcal{J}^T f_y \mathcal{J} / \mathcal{J}^T \mathcal{J})$, where T denotes transposition. Note the connection to our earlier hypothesis for boundary layer stability and, further, that the problem is dissipative, discounting errors, if $\nu(t) \leq 0$.

For the (forward or explicit) Euler method

$$y_{n+1} = y_n + h_n f(y_n, t_n),$$

the Lipschitz constant of the mapping $I + h_n f$, which might be approximated by $\|h_n f_y(y_n, t_n)\|$, should be less than 1 in order to have a contraction mapping. This exhibits the usual problem for explicit difference methods: they can become unstable if the stepsize exceeds the smallest time constant of the system.

This severe numerical instability can (sometimes) be avoided by using an *implicit* Euler method such as

$$\frac{y_{n+1} - y_n}{h_n} = f(y_{n+1}, t_{n+1})$$

(or some higher-order backward differentiation formula). When $f(y, t) = A(t)y$ is linear, we simply obtain

$$y_{n+1} = (I - A_{n+1}h_n)^{-1}y_n,$$

so iterates decrease in norm as long as $A(t)$ remains stable. More generally, for a nonlinear problem, one might use a functional iteration or substitution method such as the predictor–corrector scheme

$$y_{n+1}^{(i)} = y_n + h_n f(y_{n+1}^{(i-1)}, t_{n+1}).$$

Once again, the condition $\|h_n f_y(y, t_{n+1})\| < 1$ might be necessary to attain convergence. This would, however, again result in a severe stepsize limitation when f_y has large stable eigenvalues. An alternative consists of using the Newton-like scheme

$$y_{n+1} \simeq y_n + h_n[f(y_n, t_{n+1}) + J(y_n)(y_{n+1} - y_n)]$$

with an approximate Jacobian $J(y_n)$ which is regularly updated (say, every five time steps). Then, we would have

$$y_{n+1} = [I - h_n J(y_n)]^{-1}[y_n + h_n(f(y_n, t_{n+1}) - J(y_n)y_n)].$$

Any complication involved concerning the updating of the Jacobian should be compensated by the possibility of using longer time steps (which must still remain sufficiently small to provide accuracy).

Many times, physical systems can be written in the singularly perturbed form

$$\begin{cases} \dot{x} = f(x, y, t, \epsilon), \\ \epsilon \dot{y} = g(x, y, t, \epsilon) \end{cases}$$

(with an explicit singular perturbation parameter ϵ identified). Automatic schemes to alter the partitioning into such slow and fast subsystems during the course of integration are very much needed. Miranker (1981) provides numerical methods for such systems which are directly based on the asymptotic approximations we have already derived. Many other authors ignore the initial transient and seek only to compute the smooth limiting outer solution $\binom{X_0}{Y_0}$. This requires an initialization procedure (based on a combined asymptotic/numerical analysis) to carefully "filter" the initial conditions, i.e., to provide the appropriate initial value $Y_0(0)$. [See Kopell (1985) for an analytic treatment.]

For the singularly perturbed system, the (full) Jacobian will have the structure

$$J = \begin{pmatrix} f_x & f_y \\ \dfrac{g_x}{\epsilon} & \dfrac{g_y}{\epsilon} \end{pmatrix}.$$

Its large eigenvalues will be equal to $\mu/\epsilon + O(1)$, where μ is an eigenvalue of g_y, and its remaining eigenvalues will be equal to $\nu + O(\epsilon)$, where ν is an eigenvalue of $f_x - f_y g_y^{-1} g_x$ (the scaled Schur complement of g_y and system matrix for the variational system corresponding to the reduced problem). This suggests various iterative methods and approximation schemes, including emphasis on the dominant eigenspace corresponding to smooth solutions (and to the bounded eigenvalues) and the use of different solution procedures for different eigenspaces of the Jacobian.

The implicit Euler method takes the form

$$\begin{cases} x_{n+1} - x_n = h_n f(x_{n+1}, y_{n+1}, t_{n+1}), \\ \epsilon(y_{n+1} - y_n) = h_n g(x_{n+1}, y_{n+1}, t_{n+1}). \end{cases}$$

When $\|(\epsilon/h_n)g_y^{-1}(\ \cdot\ )\| \ll 1$, as when a relatively large stepsize is used and g_y is nonsingular, the numerical procedure mimics that of solving the reduced equation $g = 0$. Dahlquist and Söderlind (1982) emphasize the use of compound methods which treat the fast and slow parts of the system differently. For the y equations, one would seek high stability, whereas one would seek high accuracy for the x equations. It might, then, be advisable to use Newton's method for the y variables and functional iteration for the x variables.

We note that Dahlquist et al. (1982) report on practical computations. For the scalar example $\epsilon \dot{y} = y - y^3$, they observe that Newton–Raphson converges incorrectly to $Y_0 = 0$ if $0 < y(0) < 1/\sqrt{3}$, whereas a combined implicit Euler/Newton–Raphson scheme converges correctly to $Y_0 = 1$ if $y_n^2 > \frac{1}{3}(1 - \epsilon/h_n)$ is maintained.

G. Relaxation Oscillations

Consider the autonomous singularly perturbed system

$$\begin{cases} \dfrac{dx}{dt} = f(x, y), \\[2mm] \epsilon \dfrac{dy}{dt} = g(x, y) \end{cases}$$

of $m + n$ differential equations. We can expect slow motion to be primarily determined by the reduced system

$$\begin{cases} \dfrac{dx}{dt} = f(x, y), \\[2mm] 0 = g(x, y) \end{cases}$$

whereas fast motions will follow the stretched nth-order system

$$\frac{dy}{d\tau} = g(x, y)$$

[equivalently, $\epsilon\, dy/dt = g(x, y)$] with x as a parameter and $\tau = (t - t_0)/\epsilon$ for some appropriate t_0. Accordingly, slow motions will lie on the m-dimensional manifold

$$\Gamma : g(x, y) = 0,$$

whereas fast motion will occur off Γ. If the initial point $(x(0), y(0))$ is not on Γ, we might expect y to move almost instantaneously toward Γ (approximately satisfying the fast system), whereas x will hardly vary, provided the rest point $(x_1, y_1) \sim (x(0), y_1)$ for the fast system is stable [we will suppose $g_y(x(0), y_1)$ has stable eigenvalues]. Then, we can expect x and y to remain near Γ, nearly satisfying the reduced system, until g_y loses stability [as when an eigenvalue of $g_y(x_2, y_2)$ passes through zero real part]. Then, we must expect the path to move rapidly away from this bifurcation (or junction) point and away from Γ. y will then nearly satisfy the fast system whereas x will remain nearly constant until (hopefully) a stable drop point $(x_3, y_3) \sim (x_2, y_3)$ on Γ (and thereby a rest point of the fast system)

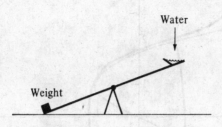

Fig. 2.6. See-saw oscillator.

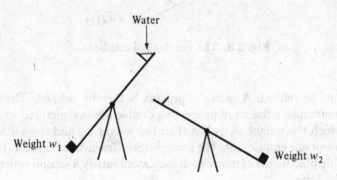

Fig. 2.7. Double see-saw oscillator.

is reached, where slow motion along Γ again begins. (We are, of course, restricting Γ in the process.) When such successive alternations between slow and fast motions produce an (asymptotically) closed trajectory, the corresponding periodic solution of our system is called a *relaxation oscillation*. It is characterized by jerky, almost instantaneous, periodic jumps in y. The period is asymptotically determined by the time spent on the slow manifold Γ where either $dt = dx/f(x,y)$ might be integrated [after $g(x,y) = 0$ was solved for $y(x)$] or $dt = [F'(y)/y]dy$ is integrated along, say, $x = F(y)$. The proper formulation of these intuitive ideas led Levinson (1951), Mishchenko (1961), and Pontryagin (1961) to define "discontinuous solutions" of the degenerate system.

Simple examples of such oscillators occur in many contexts [cf. Minorsky (1947) and Stoker (1950) for mechanical and electrical examples]. Grasman (1987) stresses biological and biochemical applications, whereas Cronin (1987) emphasizes models of electrically excitable cells. The simplest example to consider may be a see-saw with a middle pivot and a continuous stream of water flowing into a reservoir at one end. See Figure 2.6. When the weight of the water exceeds the weight on the other side, the see-saw flips, the reservoir empties, and the see-saw returns to its original

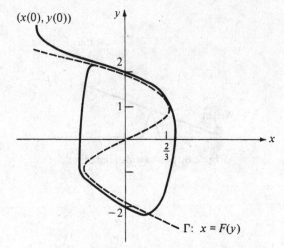

Fig. 2.8. The van der Pol oscillator.

position to be refilled. A periodic process is thereby defined. The reader should contemplate the more interesting double see-saw pictured in Figure 2.7 for which the weight w_1 is less than the weight w_2 and the water from the first reservoir empties into the second. (See Grasman's book for details).

The displacements of many such oscillators satisfy a second-order equation of the form

$$\epsilon\ddot{y} - F'(y)\dot{y} + y = 0$$

or, equivalently, after integration, the first-order system

$$\begin{cases} \dot{x} = y, \\ \epsilon\dot{y} = F(y) - x. \end{cases}$$

For the scalar example with

$$F(y) = y - \tfrac{1}{3}y^3,$$

we obtain the van der Pol oscillator (Figure 2.8) which describes (among many applications) the self-sustained oscillations of a triode circuit with a cubic current–voltage characteristic. The small ϵ then corresponds to a small self-inductance. In the x–y phase plane, the fundamental curve is $\Gamma : x = y - \tfrac{1}{3}y^3$. If $(x(0), y(0))$ lies above the upper arc Γ^+ of Γ with $x(0) < \tfrac{2}{3}$, the trajectory will drop rapidly, almost vertically, near $(x(0), (\Gamma^+)^{-1}(x(0)))$. Then, it will move to the right, remaining close to (but above) Γ until it passes the "turning point" $(\tfrac{2}{3}, 1)$ (where $g_y = F' = 1 - y^2$ loses stability); when it drops rapidly, almost vertically, to the lower arc Γ^- of Γ. Crossing Γ, it moves to the left, remaining close to, but below, Γ^- until it passes the other turning point $(-\tfrac{2}{3}, -1)$. Then, it rises rapidly, almost vertically, to the upper arc of Γ where $y \sim 2$. Crossing Γ, it follows Γ^+ to the right and continues indefinitely in nearly the same orbit.

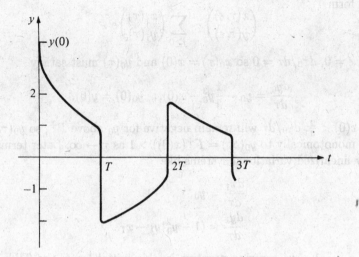

Fig. 2.9. Periodic van der Pol oscillations.

As $\epsilon \to 0$, note that the limiting (discontinuous in t) trajectory will have the period

$$T = 2 \int_{-2/3}^{2/3} \frac{dx}{y(x)} = 2 \int_{2}^{1} \frac{F'(y)}{y} dy = 3 - 2 \ln 2$$

and the y amplitude 2. Obtaining higher-order approximations to the period and amplitude remains a challenge to the applied analyst [cf. Mishchenko and Rozov (1980) and Ponzo and Wax (1965)]. Further, numerical integration of such systems poses strong challenges because the system is alternately stiff and nonstiff on and off Γ. Plotting y vs. t for $\epsilon > 0$ small, we have the nearly discontinuous solution shown in Figure 2.9. The limit lies on the stable portion of Γ, except for jumps (initially and every $T/2$ units of time thereafter).

We shall now show in more detail how the solution to this initial value problem might be constructed and how its solution approaches the limit cycle as $\epsilon \to 0$. For an initial point above the upper arc Γ^{+}, we have a transient initial layer of $O(\epsilon)$ thickness in time. If we introduce the stretched time

$$\tau = t/\epsilon,$$

we must seek a solution of the stretched system

$$\begin{cases} \dfrac{dx}{d\tau} = \epsilon y, \\ \dfrac{dy}{d\tau} = y - \dfrac{1}{3}y^3 - x \end{cases}$$

of the form

$$\begin{pmatrix} x(\tau, \epsilon) \\ y(\tau, \epsilon) \end{pmatrix} \sim \sum_{j \geq 0} \begin{pmatrix} x_j(\tau) \\ y_j(\tau) \end{pmatrix} \epsilon^j.$$

When $\epsilon = 0$, $dx_0/d\tau = 0$ so $x_0(\tau) = x(0)$ and $y_0(\tau)$ must satisfy

$$\frac{dy_0}{d\tau} = y_0 - \frac{1}{3}y_0^3 - x(0), \quad y_0(0) = y(0).$$

With $x(0) < \frac{2}{3}$, $dy_0/d\tau$ will remain negative for y_0 above Γ^+, so $y_0(\tau)$ will decay monotonically to $y_0(\infty) = \Gamma^+(x(0)) > 1$ as $\tau \to \infty$. Later terms will satisfy linearized variational systems like

$$\frac{dx_1}{d\tau} = y_0,$$

$$\frac{dy_1}{d\tau} = (1 - y_0^2)y_1 - x_1.$$

Since $g_y := 1 - y^2$ remains negative in this initial layer, the Tikhonov–Levinson theory will guarantee that the solution matches an outer solution $(X(t, \epsilon), Y(t, \epsilon))$ as $\tau \to \infty$ [i.e., for some $t_0 = O(\epsilon)$ which could be determined accurately by integrating $d\tau$].

Subsequently, the limiting outer solution $(X_0(t), Y_0(t))$ will satisfy the reduced system

$$\begin{cases} \dot{X}_0 = Y_0, \\ 0 = Y_0 - \frac{1}{3}Y_0^3 - X_0. \end{cases}$$

Moreover, Y_0 will be uniquely determined as a function of X_0 by the "initial" value $(X_0(0^+), Y_0(0^+)) = (x(0), y_0(\infty))$ on Γ^+, as long as the Jacobian $1 - Y_0^2$ remains negative. More explicitly, we could integrate $(1 - Y_0^2)\dot{Y}_0 = \dot{X}_0 = Y_0$ to obtain $\ln Y_0 - \frac{1}{2}Y_0^2 = t + C_0$ where C_0 is determined by matching with the initial layer since $Y_0(0^+)$ is known. Thus, we also obtain $X_0 = Y_0 - \frac{1}{3}Y_0^3$. Later terms in this outer expansion will then follow from linear systems like

$$\dot{X}_1 = Y_1,$$

$$\dot{Y}_0 = (1 - Y_0^2)Y_1 - X_1.$$

Thus, $Y_1 = (\dot{Y}_0 + X_1)/(1 - Y_0^2)$ and there remains a linear initial value problem for X_1. The singularities near $Y_0 = 1$ (i.e., near $t_1 = -C_0 - \frac{1}{2}$) could be analyzed more explicitly.

To get around $y = 1$, we attempt a new local stretching

$$\begin{cases} t = t_1 + \epsilon^\alpha \zeta, \\ x = \frac{2}{3} + \epsilon^\beta \xi, \\ y = 1 + \epsilon^\gamma \eta, \end{cases}$$

selecting the positive constants $\alpha, \beta,$ and γ to achieve a meaningful limiting problem. Since the system for x and y then implies

$$\epsilon^{\beta-\alpha}\frac{d\xi}{d\zeta} = 1 + \epsilon^{\gamma}\eta$$

and

$$\epsilon^{1+\gamma-\alpha}\frac{d\eta}{d\zeta} = 1 + \epsilon^{\gamma}\eta - \tfrac{1}{3}(1+\epsilon^{\gamma}\eta)^3 - \left(\frac{2}{3}+\epsilon^{\beta}\xi\right)$$

$$= -\epsilon^{2\gamma}\eta^2 - \frac{1}{3}\epsilon^{3\gamma}\eta^3 - \epsilon^{\beta}\xi,$$

we will retain the maximum number of terms as $\epsilon \to 0$ if we set $\beta = \alpha$ and $1 + \gamma - \alpha = 2\gamma = \beta$. Thus, we introduce the stretched variable

$$\zeta = \frac{t - t_1}{\epsilon^{2/3}}$$

and we naturally seek a solution of

$$\begin{cases} \dfrac{d\xi}{d\zeta} = 1 + \epsilon^{1/3}\eta, \\[2mm] \dfrac{d\eta}{d\zeta} = -\eta^2 - \xi - \dfrac{1}{3}\epsilon^{1/3}\eta^3 \end{cases}$$

as a power series in $\epsilon^{1/3}$ which matches the outer solution along Γ^+ as $\zeta \to -\infty$. Leading terms imply that

$$\xi_0(\zeta) = \zeta + k_0$$

and

$$\frac{d\eta_0}{d\zeta} = -\eta_0^2 - \zeta - k_0,$$

whereas later terms will satisfy corresponding variational problems. The latter Riccati equation can be solved by setting

$$\eta_0(\zeta) = \frac{z'(\zeta)}{z(\zeta)}$$

where z satisfies the Airy equation

$$z'' + (\zeta + k_0)z = 0.$$

In terms of the modified Bessel functions K_j and I_j with imaginary arguments [cf. Abramovitz and Stegen (1965)], we must have

$$z(\zeta) = (k_0 - \zeta)^{1/2}\left\{ AK_{1/3}\left(\tfrac{2}{3}(k_0 - \zeta)^{3/2}\right) + BI_{1/3}\left(\tfrac{2}{3}(k_0 - \zeta)^{3/2}\right)\right\}$$

for constants A and B. Alternative representations in terms of the Airy functions Ai and Bi are also helpful. Matching with the preceding outer solution as $\zeta \to -\infty$ requires us to take $B = 0$ and $k_0 = 0$, so η_0 is known. It will, however, have a simple pole at the first zero of $z(\zeta)$, i.e., when $\tfrac{2}{3}\zeta_2^{3/2} \approx 2.383$ and $t_2 = t_1 + \epsilon^{2/3}\zeta_2$. Later terms will also be singular there.

Due to the singularity, we try a new stretching about t_2 to attempt to get from $y_2 = 1 + \epsilon^{1/3}\eta(\zeta_2)$ to the lower arc of Γ where $y_2 \approx -2$. Using the familiar stretching

$$\kappa = (t - t_2)/\epsilon,$$

we must integrate (as on the first fast interval)

$$\begin{cases} \dfrac{dx}{d\kappa} = \epsilon y, \\ \dfrac{dy}{d\kappa} = y - \dfrac{1}{3}y^3 - x. \end{cases}$$

The system suggests that it would be natural to take $\binom{x}{y}$ as a power series in ϵ, but matching requires the coefficients to be ϵ-dependent. Taking $x_0(\kappa) = \frac{2}{3} + \zeta_2\epsilon^{2/3}$, we will have $dy_0/d\kappa = y_0 - \frac{1}{3}y_0^3 - \frac{2}{3} - \zeta_2\epsilon^{2/3}$. Following arguments of Carrier (1974), we rewrite this as

$$\frac{dy_0}{d\kappa} = -\frac{1}{3}(y_0 - 1)^2 \left(y_0 + 2 + \frac{1}{3}\zeta_2\epsilon^{2/3} \right) + O(\epsilon^{2/3}).$$

This implies a monotonic relation between y_0 and κ until $y_3 \approx -2 - \frac{1}{3}\zeta_2\epsilon^{2/3}$ and forces a final stretching with

$$u = \frac{y - y_3}{\epsilon} \quad \text{and} \quad \sigma = \frac{1}{\epsilon}\left(t - t_2 + \frac{1}{3}\epsilon \ln \epsilon \right).$$

Matching with the outer solution near the lower arc of Γ provides fairly complicated expansions for the amplitude and period of the limiting "limit cycle." In particular, note that the asymptotic sequences involved are not easy to predict a priori.

This approach, like that of Carrier (1953, 1974), seeks a simple asymptotic solution, which is continued in a clever way each time it breaks down. A much more complete matching is carried out in Nipp (1988) which also analyzes a third-order system, the Field–Noyes model of the chemical oscillations of the Belousov–Zhabotinskii reaction, by using *fifteen* successive stretchings and matchings! MacGillivray (1990) also provides valuable reading.

Exercise

[Compare Stoker (1950)]

Consider the linear relaxation oscillator determined by using the piecewise linear characteristic function

$$F(y) = \begin{cases} -2 - y, & y < -1, \\ y, & -1 \le y \le 1, \\ 2 - y, & y > 1. \end{cases}$$

Show that the resulting oscillation has the period

$$T = 2\log 3 - \tfrac{8}{3}\epsilon\log\epsilon + \epsilon\left(\tfrac{8}{3} - 2\log 3\right) + \cdots.$$

H. A Combustion Model

Reiss (1980) introduced, as a simple combustion model, the initial value problem

$$\dot{y} = y^2(1 - y), \qquad y(0) = \epsilon$$

on $t \geq 0$. Here y represents the concentration of the reacting chemical at time t and $\epsilon > 0$ gives a small disturbance of the preignition state $y = 0$. The exact solution $y(t, \epsilon)$ is easily obtained (for example, by setting $w = 1/y$). It has the implicit form

$$\frac{1}{y} + \ln\left(\frac{1}{y} - 1\right) = \frac{1}{\epsilon} + \ln\left(\frac{1}{\epsilon} - 1\right) - t.$$

Its asymptotic behavior as $\epsilon \to 0$ is not obvious. Note, however, that the sign of $\dot{y}$ implies that y will increase monotonically with t to its explosive steady state $y = 1$. Careful numerical integration shows that the solution remains small until t reaches about $1/\epsilon$, where it increases rapidly to the final explosive state. Moreover, the smaller the value of ϵ, the longer the solution stays near the preignition state and the more rapid is the ultimate move to the explosive state $y = 1$. (See Figure 2.10.)

Our presentation of the asymptotic solution is largely based on Kassoy (1982) and Kapila (1989). We shall first seek an outer solution

$$y(t, \epsilon) = \epsilon Y(t, \epsilon) = \epsilon(Y_0(t) + \epsilon Y_1(t) + \epsilon^2 Y_2(t) + \cdots),$$

scaled by the initial value. Equating coefficients, then, in the resulting initial value problem

$$\dot{Y} = \epsilon Y^2(1 - \epsilon Y), \qquad Y(0, \epsilon) = 1,$$

we ask that

$$\dot{Y}_0 = 0, \qquad Y_0(0) = 1,$$

$$\dot{Y}_1 = Y_0^2, \qquad Y_1(0) = 0,$$

$$\dot{Y}_2 = 2Y_0Y_1 - Y_0^3, \quad Y_2(0) = 0,$$

etc. The resulting outer solution

$$y(t, \epsilon) = \epsilon + \epsilon^2 t + \epsilon^3(t^2 - t) + \cdots$$

therefore provides the small preignition solution asymptotically. Since each Y_j grows like t^j as $t \to \infty$, we can anticipate breakdown of this solution when

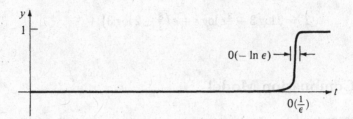

Fig. 2.10. Explosion model.

$$\lambda = \epsilon t = O(1),$$

because the terms of the series then all have the same asymptotic size. This could, of course, be verified by an asymptotic expansion of the exact solution.

The breakdown suggests an alternative outer expansion

$$y(t, \epsilon) = \epsilon Z(\lambda, \epsilon) = \epsilon (Z_0(\lambda) + \epsilon Z_1(\lambda) + \epsilon^2 Z_2(\lambda) + \cdots)$$

expressed in terms of the slower time $\lambda = \epsilon t$. Equating coefficients in

$$\frac{dZ}{d\lambda} = Z^2(1 - \epsilon Z), \qquad Z(0) = 1$$

requires

$$\frac{dZ_0}{d\lambda} = Z_0^2, \qquad Z_0(0) = 1,$$

$$\frac{dZ_1}{d\lambda} = 2Z_0 Z_1 - Z_0^2 Z_1, \qquad Z_1(0) = 0,$$

$$\frac{dZ_2}{d\lambda} = 2Z_0 Z_2 + Z_1^2 - 2Z_0^2 Z_1 - 2Z_0 Z_1^2, \quad Z_2(0) = 0,$$

etc. This yields the outer solution

$$\epsilon Z(\lambda, \epsilon) = \frac{\epsilon}{1 - \lambda} + \frac{\epsilon^2 \ln(1 - \lambda)}{(1 - \lambda)^2} + \epsilon^3 \left(\frac{\ln^2(1 - \lambda)}{(1 - \lambda)^3} \right.$$

$$\left. - \frac{\ln(1 - \lambda)}{(1 - \lambda)^3} - \frac{1}{(1 - \lambda)^3} + \frac{1}{(1 - \lambda)^2} \right) + \cdots$$

which, clearly, includes the preceding outer solution when $t = \lambda/\epsilon$ is finite. This expansion certainly breaks down as $\lambda \to 1^-$, i.e., as $t \to (1/\epsilon)^-$. In particular, we have trouble when $[\epsilon \ln(1 - \lambda)]/(1 - \lambda) = O(1)$. Thus, the regular perturbation process ceases to be valid, and nonuniform convergence occurs, near $t = \infty$.

The preceding analysis might suggest introducing a final new time scale

$$\mu = \frac{\lambda - 1 + \epsilon\tilde{\mu}(\epsilon)}{\epsilon}$$

where the time shift $\epsilon\tilde{\mu}(\epsilon)$ is to be determined to obtain a solution

$$y(t, \epsilon) = \nu(\mu, \epsilon)$$

which matches $\epsilon Z(\lambda, \epsilon)$ as $\mu \to -\infty$. [We will have $\tilde{\mu}(\epsilon)$ unbounded as $\epsilon \to 0$, but very mildly so.] Since ν must satisfy the parameter-free equation

$$\frac{d\nu}{d\mu} = \nu^2(1 - \nu),$$

we can obtain a solution $\nu(\mu)$, independent of ϵ, if the shift $\tilde{\mu}(\epsilon)$ is selected appropriately. Specifically, the transition-layer solution

$$-\frac{1}{\nu} + \ln\left(\frac{\nu}{1 - \nu}\right) = \mu$$

matches $\nu = 1$ as $\mu \to \infty$, so the explosive state will be achieved. Expanding about $\mu = -\infty$, we will have

$$\nu = -\frac{1}{\mu} - \frac{1}{\mu^2}\ln\left(-\frac{1}{\mu}\right) - \cdots.$$

Substituting the power series expansion in ϵ for

$$-\frac{1}{\mu} = \frac{\epsilon}{1 - \lambda}\left(1 - \frac{\epsilon\tilde{\mu}}{1 - \lambda}\right)^{-1},$$

we obtain

$$\nu \sim \frac{\epsilon}{1 - \lambda} + \frac{\epsilon^2}{(1 - \lambda)^2}[\ln(1 - \lambda) - \tilde{\mu} - \ln\epsilon] + \cdots.$$

This matches the solution ϵZ to $O(\epsilon^2)$ if we pick

$$\tilde{\mu}(\epsilon) \sim -\ln\epsilon.$$

Later terms in the asymptotic expansion (as well as the exact solution) show that $\tilde{\mu}(\epsilon)$ must actually be asymptotically equal to $-\ln\epsilon + \ln(1 - \epsilon)$. This shows that the solution jumps from $y = 0$ to $y = 1$ in a transition layer of $O(-\ln\epsilon)$ thickness about $t = 1/\epsilon$. The introduction of such logarithmic terms (or other similar intermediate-order terms) is common in practice and is sometimes referred to as transcendental switchback [cf. Lagerstrom (1988)]. For more realistic combustion problems, the reader is referred to Kapila (1983), Dold (1985), and Buckmaster and Ludford (1982).

I. Linear and Nonlinear Examples of Conditionally Stable Systems

In our earlier study of initial value problems for vector systems

$$\begin{cases} \dot{x} = f(x, y, t, \epsilon), \\ \epsilon \dot{y} = g(x, y, t, \epsilon), \end{cases}$$

it was quite critical (for stability within the initial layer and along the limiting outer solution) that the Jacobian matrix g_y remained stable in appropriate domains. In many practical problems (including some arising in control theory), we encounter situations where g_y has unstable and/or neutrally stable eigenvalues in addition to stable ones. Let us restrict attention here to hyperbolic cases where g_y has $k < n$ eigenvalues which are strictly in the left half-plane and $n - k$ eigenvalues strictly in the right half-plane. According to the center manifold theorem [cf. Kelley (1967) and Carr (1981)], the autonomous limiting boundary layer system

$$\frac{dz_0}{d\tau} = g(x(0), z_0, 0, 0), z_0(0) = y(0)$$

will [at least sufficiently near $z_0(\infty)$] have a decaying solution $z_0(\tau)$ if $y_0(0)$ is restricted to an appropriate k-dimensional stable initial manifold S_0. (Near the rest point, S_0 could be approximated using the local linearization of g.) Since the associated variational system can be analyzed (in terms of a k-dimensional subspace of the full n space), a resulting k-dimensional stable manifold for the complete inner problem can be completely described asymptotically; i.e., we are able to obtain an asymptotic expansion $S(\epsilon) \sim \sum_{j \geq 0} S_j \epsilon^j$ for the stable initial manifold for $y(0, \epsilon)$ [cf. Levin and Levinson (1954), Levin (1957), and Hoppensteadt (1971)]. If $y(0, \epsilon)$ lies in $S(\epsilon)$, we can expect to find an asymptotic solution of our initial value problem (as in the strictly stable case) in the form

$$\begin{cases} x(t, \epsilon) = X(t, \epsilon) + \epsilon \xi(\tau, \epsilon), \\ y(t, \epsilon) = Y(t, \epsilon) + \eta(\tau, \epsilon), \end{cases}$$

where the boundary layer correction $(\epsilon\xi, \eta) \to 0$ as $\tau = t/\epsilon \to \infty$. If $y(0, \epsilon)$ lies off $S(\epsilon)$, we must instead expect blowup of y as $\epsilon \to 0$ for $t > 0$.

(i)   A linear problem.
Consider the initial value problem for

$$\begin{cases} \dot{x} = A(t, \epsilon)x + B(t, \epsilon)y + C(t, \epsilon), \\ \epsilon \dot{y} = D(t, \epsilon)x + E(t, \epsilon)y + F(t, \epsilon) \end{cases}$$

on $0 \leq t \leq 1$ where $E(t, 0)$ is assumed to have k stable and $n - k$ unstable eigenvalues. We will assume that all coefficients are infinitely differentiable with respect to t and that they and the initial vectors $x(0, \epsilon)$ and $y(0, \epsilon)$ have asymptotic power series expansions in ϵ. Let us assume the factorization

$$E(t, 0) = S(t) \begin{pmatrix} Q(t) & 0 \\ 0 & P(t) \end{pmatrix} S^{-1}(t)$$

to block diagonal form, where $Q(t)$ is a strictly stable $k \times k$ matrix and $P(t)$ is unstable. This could be obtained numerically [even with $S(t)$ orthogonal] by use of Riccati transformations [cf. Flaherty and O'Malley (1980)]. Moreover, the invertible change of variables

$$y = S(t)w$$

transforms our second equation to

$$\epsilon \dot{w} = S^{-1}(t)[D(t, \epsilon)x + (E(t, \epsilon)S(t) - \epsilon \dot{S}(t))w + F(t, \epsilon)].$$

Partitioning w as $\binom{w_1}{w_2}$ for a k vector w_1, we obtain a system in the form

$$\begin{cases} \dot{x} = A(t, \epsilon)x + B_1(t, \epsilon)w_1 + B_2(t, \epsilon)w_2 + C(t, \epsilon), \\ \epsilon \dot{w}_1 = D_1(t, \epsilon)x + (Q(t) + \epsilon R_{11}(t, \epsilon))w_1 \\ \qquad\qquad + \epsilon R_{12}(t, \epsilon)w_2 + F_1(t, \epsilon), \\ \epsilon \dot{w}_2 = D_2(t, \epsilon)x + \epsilon R_{21}(t, \epsilon)w_1 \\ \qquad\qquad + [P(t) + \epsilon R_{22}(t, \epsilon)]w_2 + F_2(t, \epsilon) \end{cases}$$

since $S^{-1}(t)E(t, 0)S(t)$ is block diagonal. We shall now show how to construct a family of asymptotic solutions of the form

$$\begin{pmatrix} x(t, \epsilon) \\ w_1(t, \epsilon) \\ w_2(t, \epsilon) \end{pmatrix} = \begin{pmatrix} X(t, \epsilon) \\ W_1(t, \epsilon) \\ W_2(t, \epsilon) \end{pmatrix} + \begin{pmatrix} \epsilon \xi(\tau, \epsilon) \\ \eta_1(\tau, \epsilon) \\ \eta_2(\tau, \epsilon), \end{pmatrix},$$

where ξ, η_1, and $\eta_2 \to 0$ as $\tau = t/\epsilon \to \infty$. These solutions will be completely specified by the initial vectors $x(0, \epsilon)$ and $w_1(0, \epsilon)$. The $(n - k)$-dimensional vector $w_2(0, \epsilon)$ is not free. This implies that $y(t, \epsilon)$ will have the anticipated asymptotic structure, with $y(0, \epsilon)$ restricted to a k manifold which we will describe asymptotically as $\mathcal{S}(\epsilon)$.

The outer solution

$$\begin{pmatrix} X \\ W_1 \\ W_2 \end{pmatrix}$$

must obviously satisfy the system

$$
\begin{cases}
\dot{X} = A(t,\epsilon)X + B_1(t,\epsilon)W_1 + B_2(t,\epsilon)W_2 + C(t,\epsilon), \\[2mm]
\epsilon\dot{W}_1 = D_1(t,\epsilon)X + (Q(t) + \epsilon R_{11}(t,\epsilon))W_1 \\
\qquad\quad + \epsilon R_{12}(t,\epsilon)W_2 + F_1(t,\epsilon), \\[2mm]
\epsilon\dot{W}_2 = D_2(t,\epsilon)X + \epsilon R_{21}(t,\epsilon)W_1 \\
\qquad\quad + [P(t) + \epsilon R_{22}(t,\epsilon)]W_2 + F_2(t,\epsilon)
\end{cases}
$$

as a power series in ϵ. Thus, the limiting problem will have the solution

$$
W_{10} = -Q^{-1}(t)[D_{10}(t)X_0 + F_{10}(t)]
$$

and

$$
W_{20} = -P^{-1}(t)[D_{20}(t)X_0 + F_{20}(t)],
$$

where X_0 satisfies the reduced-order initial value problem

$$
\dot{X}_0 = (A_0 - B_{10}Q^{-1}D_{10} - B_{20}P^{-1}D_{20})X_0 \\
\quad + (C_0 - B_{10}Q^{-1}F_{10} - B_{20}P^{-1}F_{20}), \quad X_0(0) = x(0,0) \equiv x_{00}.
$$

Existence and uniqueness of X_0 on $0 \le t \le 1$ follows from the smoothness of the coefficients. Later terms in the outer expansions are likewise uniquely determined by the, yet unspecified, initial vectors $X_j(0)$. Thus, the $(m+n)$-dimensional system has an m-dimensional manifold of smooth solutions parameterized by $X(0,\epsilon)$. The initial layer correction

$$
\begin{pmatrix} \epsilon\xi \\ \eta_1 \\ \eta_2 \end{pmatrix}
$$

must satisfy the stretched homogeneous system

$$
\begin{cases}
\dfrac{d\xi}{d\tau} = \epsilon A(\epsilon\tau,\epsilon)\xi + B_1(\epsilon\tau,\epsilon)\eta_1 + B_2(\epsilon\tau,\epsilon)\eta_2, \\[3mm]
\dfrac{d\eta_1}{d\tau} = \epsilon D_1(\epsilon\tau,\epsilon)\xi + (Q(\epsilon\tau) + \epsilon R_{11}(\epsilon\tau,\epsilon))\eta_1 \\
\qquad\qquad + \epsilon R_{12}(\epsilon\tau,\epsilon)\eta_2, \\[3mm]
\dfrac{d\eta_2}{d\tau} = \epsilon D_2(\epsilon\tau,\epsilon)\xi + \epsilon R_{21}(\epsilon\tau,\epsilon)\eta_1 \\
\qquad\qquad + [P(\epsilon\tau) + \epsilon R_{22}(\epsilon\tau,\epsilon)]\eta_2
\end{cases}
$$

as a power series in ϵ whose terms must decay to zero as $\tau \to \infty$. Thus, the leading terms will satisfy the constant system

$$
\frac{d}{d\tau}\begin{pmatrix} \xi_0 \\ \eta_{10} \\ \eta_{20} \end{pmatrix} = \begin{pmatrix} 0 & B_{10}(0) & B_{20}(0) \\ 0 & Q(0) & 0 \\ 0 & 0 & P(0) \end{pmatrix} \begin{pmatrix} \xi_0 \\ \eta_{10} \\ \eta_{20} \end{pmatrix}.
$$

To achieve decay, then, as $\tau \to \infty$, the stability of the matrix $Q(0)$ and the instability of $P(0)$ imply that we must have

$$
\begin{cases}
\eta_{10}(\tau) = e^{Q(0)\tau}\eta_{10}(0), \\[2mm]
\eta_{20}(\tau) = 0, \\[2mm]
\xi_0(\tau) = B_{10}(0)Q^{-1}(0)e^{Q(0)\tau}\eta_{10}(0).
\end{cases}
$$

The k vector $\eta_{10}(0)$ is still arbitrary, but $\eta_{20}(0) = 0$ restricts $y(0,0)$ to the k manifold S_0 defined by

$$
y(0,0) = S(0)\begin{pmatrix} W_{10}(0) + \eta_{10}(0) \\ W_{20}(0) \end{pmatrix}.
$$

Having determined $\xi_0(0)$ as a function of $\eta_{10}(0)$, we can uniquely determine the next term X_1 in the outer expansion by integrating its linear differential equation using the initial value

$$
X_1(0) = x_{01} - B_{10}(0)Q^{-1}(0)\eta_{10}(0).
$$

Later terms in the initial layer correction must satisfy the analogous non-homogenous system with exponentially decaying forcing terms. From the coefficient of ϵ^1, for example, we obtain

$$
\begin{cases}
\dfrac{d\xi_1}{d\tau} = B_{10}(0)\eta_{11} + B_{20}(0)\eta_{21} + A_0(0)\xi_0 + \left(B_{11}(0) + \tau\dfrac{dB_{10}}{dt}(0) \right)\eta_{10}, \\[3mm]
\dfrac{d\eta_{11}}{d\tau} = Q(0)\eta_{11} + D_{10}(0)\xi_0 + \left(\tau\dfrac{dQ(0)}{dt} + R_{110}(0) \right)\eta_{10}, \\[3mm]
\dfrac{d\eta_{21}}{d\tau} = P(0)\eta_{21} + D_{20}(0)\xi_0 + R_{210}(0)\eta_{10}.
\end{cases}
$$

Thus, the decaying solutions are given by

$$
\eta_{11}(\tau) = e^{Q(0)\tau}\eta_{11}(0) + \left(\int_0^\tau e^{Q(0)(\tau-s)}[D_{10}(0)B_{10}(0)Q^{-1}(0) \right.
$$

$$
\left. + s\dfrac{dQ(0)}{dt} + R_{110}(0)]e^{Q(0)s}ds \right)\eta_{10}(0),
$$

$$
\eta_{21}(\tau) = -\left(\int_\tau^\infty e^{P(0)(\tau-r)}[D_{20}(0)B_{10}(0)Q^{-1}(0) + R_{210}(0)]e^{Q(0)r}dr \right)\eta_{10}(0),
$$

and

$$\xi_1(\tau) = -\int_\tau^\infty [B_{10}(0)\eta_{11}(r) + B_{20}(0)\eta_{21}(r) + A_0(0)\xi_0(r)$$

$$+ \left(B_{11}(0) + r\frac{dB_{20}}{dt}(0)\right)\eta_{10}(r)]dr$$

with only $\eta_{11}(0)$ free. (Any other solutions will not decay appropriately.) Thus, it follows that the initial value $y_1(0)$ of the $O(\epsilon)$ term in the asymptotic expansion for $y(0, \epsilon)$ is specified through the two k manifolds S_0 and S_1. Moreover, the initial value $\xi_1(0)$ determines $X_2(0) = x_{02} - \xi_1(0)$ and allows us to completely obtain the second-order terms of the outer expansion. Approximations to the original y variable follow immediately.

We note that it would be equivalent, but somewhat more complicated, to match inner and outer solutions instead of seeking the initial layer correction. An advantage of the present procedure is that we have directly determined a uniformly valid asymptotic approximation instead of separate approximations in the inner and outer regions.

(ii)   A nonlinear example [cf. Jeffries and Smith (1989)].
    Let us consider the initial value problem

$$\begin{cases} \dot{x} = -y_2^2, & x(0, \epsilon) = 1, \\ \epsilon\dot{y}_1 = y_1^2 - x^2, & y_1(0, \epsilon) \text{ given,} \\ \epsilon\dot{y}_2 = -2y_1 + 2y_2, & y_2(0, \epsilon) \text{ to be determined} \end{cases}$$

and let us seek a bounded asymptotic solution on the interval $0 \le t \le 1$. The example is artificial, but similar problems occur in authentic applications. It provides an example of some more complicated matching. Here, the reduced problem

$$\begin{cases} \dfrac{dX_0}{dt} = -Y_{20}^2, & X_0(0) = 1, \\ 0 = Y_{10}^2 - X_0^2, \\ 0 = -2Y_{10} + 2Y_{20} \end{cases}$$

has the two solutions

$$\begin{pmatrix} X_0 \\ Y_{10} \\ Y_{20} \end{pmatrix}_\pm = \begin{pmatrix} 1 \\ \pm 1 \\ \pm 1 \end{pmatrix}\left(\frac{1}{1+t}\right).$$

Moreover, the Jacobian $g_y = \begin{pmatrix} 2y_1 & 0 \\ -2 & 2 \end{pmatrix}$ of the fast system has two unstable eigenvalues when $y_1 > 0$ and one stable and one unstable eigenvalue when $y_1 < 0$. [The preceding theory implies that the limit

$$\begin{pmatrix} X_0 \\ Y_{10} \\ Y_{20} \end{pmatrix}_+$$

would be appropriate for a terminal value problem on any interval $-1 < -T \leq t \leq 0$. (The interval has to be restricted so that the outer solution $X_0(t) = 1/(1+t)$ does not blow up.) There would then need to be an $O(\epsilon)$ thick terminal layer, provided $y_1(0,0) > -1$, and $y_2(0,\epsilon)$ could be left arbitrary.] For our initial value problem, however, we will show that the limit

$$\begin{pmatrix} X_0 \\ Y_{10} \\ Y_{20} \end{pmatrix}_-$$

is appropriate on any finite interval $0 \leq t \leq T$ provided $y_1(0,0) < 1$ and $y_2(0,\epsilon)$ is restricted to a stable manifold which we will construct asymptotically via a power series expansion approach.

The inner problem corresponding to the stretching $\tau = t/\epsilon$ is given by

$$\begin{cases} \dfrac{du}{d\tau} = -\epsilon v_2^2, & u(0,\epsilon) = 1, \\[2mm] \dfrac{dv_1}{d\tau} = v_1^2 - u^2, & v_1(0,\epsilon) = y_1(0,\epsilon), \\[2mm] \dfrac{dv_2}{d\tau} = -2v_1 + 2v_2, & v_2(0,\epsilon) = y_2(0,\epsilon). \end{cases}$$

For convenience, we will expand the initial vectors as

$$y_j(0,\epsilon) \sim \sum_{k=0}^{\infty} y_{jk}^0 \epsilon^k, \quad j = 1, 2.$$

Thus, the limiting inner problem will be given by

$$\begin{cases} \dfrac{du_0}{d\tau} = 0, & u_0(0) = 1, \\[2mm] \dfrac{dv_{10}}{d\tau} = v_{10}^2 - u_0^2, & v_{10}(0) = y_{10}^0, \\[2mm] \dfrac{dv_{20}}{d\tau} = -2v_{10} + 2v_{20}, & v_{20}(0) = y_{20}^0. \end{cases}$$

Integrating the first two equations, we obtain

$$u_0(\tau) = 1$$

and

$$v_{10}(\tau) = -1 + \frac{2e^{-2\tau}}{k + e^{-2\tau}},$$

where $k = (1 - y_{10}^0)/(1 + y_{10}^0)$ and we will make the restriction $y_{10}^0 < 1$ in order that $v_{10}(\tau)$ stays bounded for all $\tau \geq 0$. Then $v_{10} = -1 + O(e^{-2\tau})$, so the equation for v_{20} can be rewritten as

$$\frac{d}{d\tau}(v_{20} + 1) = -2(v_{10} + 1) + 2(v_{20} + 1).$$

[This rewriting can be motivated by the anticipated matching to the outer limit

$$\begin{pmatrix} 1 \\ -1 \\ -1 \end{pmatrix}$$

as $\tau \to \infty$.] Integrating from infinity, then, provides the particular solution

$$v_{20}^p(\tau) = -1 + 4 \int_\tau^\infty \frac{e^{-2s} e^{2(\tau - s)}}{k + e^{-2s}} ds$$

$$= -1 + 2 \ln \left(\frac{e}{(1 + \frac{1}{ke^{2\tau}}) k e^{2\tau}} \right) = -1 + O(e^{-2\tau}).$$

Since the complementary solution $e^{2\tau} c$ becomes exponentially unbounded for any $c \neq 0$, we take v_{20} to be v_{20}^p. Then, the initial value $y_2(0, \epsilon)$ must be restricted so that

$$y_{20}^0 = -1 + 2 \ln \left(\frac{e}{(1 + 1/k)^k} \right) = 1 - 2 \frac{(1 - y_{10}^0)}{(1 + y_{10}^0)} \ln \left(\frac{2}{1 - y_{10}^0} \right).$$

The initial values y_{2j}^0 for later terms in the expansion for $y_2(0, \epsilon)$ will have to be analogously determined.

From the $O(\epsilon)$ terms in the inner expansion, we obtain the linear system

$$\begin{cases} \dfrac{du_1}{d\tau} = -v_{20}^2, & u_1(0) = 0, \\[2mm] \dfrac{dv_{11}}{d\tau} = 2v_{10}v_{11} - 2u_1, & v_{11}(0) = y_{11}^0, \\[2mm] \dfrac{dv_{21}}{d\tau} = 2v_{21} - 2v_{11}, & v_{21}(0) = y_{21}^0. \end{cases}$$

Thus,

$$u_1(\tau) = -\int_0^\tau v_{20}^2(s) ds = -\tau - \alpha + \int_\tau^\infty (v_{20}^2 - 1) ds = -\tau - \alpha + O(e^{-2\tau}),$$

where $\alpha = \int_0^\infty (v_{20}^2 - 1) ds$. Rewriting the equation for v_{11} in the suggestive form

$$\frac{d}{d\tau}\left(v_{11} - \tau - \alpha + \tfrac{1}{2}\right) = 2v_{10}\left(v_{11} - \tau - \alpha + \tfrac{1}{2}\right) - 2(u_1 + \tau + \alpha)$$

$$+ (v_{10} + 1)(2\tau + 2\alpha - 1)$$

and integrating yields

$$v_{11} = \tau + \alpha - \tfrac{1}{2} + e^{2\int_0^\tau v_{10}(s)ds}\left(y_{11}^o - \alpha + \tfrac{1}{2}\right) - 2\int_0^\tau e^{2\int_r^\tau v_{10}(s)ds} O(1+r)e^{-2r}dr$$

so

$$v_{11} = \tau + \alpha - \tfrac{1}{2} + O(e^{-2\tau}).$$

Finally, the equation for v_{21} implies that

$$\frac{d}{d\tau}(v_{21} - \tau - \alpha) = 2(v_{21} - \tau - \alpha) - 2\left(v_{11} - \tau - \alpha + \tfrac{1}{2}\right).$$

Thus, the only possible solution v_{21} which does not grow exponentially is

$$v_{21}(\tau) = \tau + \alpha + 2\int_\tau^\infty e^{2(\tau - s)}\left[v_{11}(s) - s - \alpha + \tfrac{1}{2}\right]ds = \tau + \alpha + O(e^{-2\tau}).$$

Note that this restricts $y_2(0, \epsilon)$ such that

$$y_{21}^o = \alpha + 2\int_0^\infty e^{-2s}\left[v_{11}(s) - s - \alpha + \tfrac{1}{2}\right]ds.$$

Moreover, we have shown that the inner solution is of the form

$$\begin{pmatrix} u(\tau, \epsilon) \\ v_1(\tau, \epsilon) \\ v_2(\tau, \epsilon) \end{pmatrix} \sim \begin{pmatrix} 1 \\ -1 \\ -1 \end{pmatrix} + \epsilon \begin{pmatrix} -\tau - \alpha \\ \tau + \alpha - \tfrac{1}{2} \\ \tau + \alpha \end{pmatrix} + O(\epsilon^2) + O(e^{-2\tau}).$$

This should match the outer solution which, as a function of τ, takes the form

$$\begin{pmatrix} X(\epsilon\tau, \epsilon) \\ Y_1(\epsilon\tau, \epsilon) \\ Y_2(\epsilon\tau, \epsilon) \end{pmatrix} \sim \begin{pmatrix} X_0(\epsilon\tau) + \epsilon X_1(0) \\ Y_{10}(\epsilon\tau) + \epsilon Y_{11}(0) \\ Y_{20}(\epsilon\tau) + \epsilon Y_{21}(0) \end{pmatrix} + O(\epsilon^2)$$

$$\sim \begin{pmatrix} X_0(0) \\ Y_{10}(0) \\ Y_{20}(0) \end{pmatrix} + \epsilon \begin{pmatrix} \tau X_0'(0) + X_1(0) \\ \tau Y_{10}'(0) + Y_{11}(0) \\ \tau Y_{20}'(0) + Y_{21}(0) \end{pmatrix} + O(\epsilon^2).$$

Thus, matching requires $X_0(0) = 1$ and $Y_{10}(0) = -1$, as anticipated, and the limiting solution for $t > 0$ will be given by

$$\begin{pmatrix} X_0(t) \\ Y_{10}(t) \\ Y_{20}(t) \end{pmatrix} = \begin{pmatrix} 1 \\ -1 \\ -1 \end{pmatrix} \frac{1}{1+t}.$$

Expanding $1/(1+t) = 1 - \epsilon\tau + \epsilon^2\tau^2 - \cdots$ suggests that polynomials in τ should occur in the matching process. Further, since

$$\begin{pmatrix} X_0'(0) \\ Y_{10}'(0) \\ Y_{20}'(0) \end{pmatrix} = \begin{pmatrix} -1 \\ 1 \\ 1 \end{pmatrix},$$

the $O(\epsilon)$ terms in the inner and outer expansion will match for large τ if we select

$$\begin{pmatrix} X_1(0) \\ Y_{11}(0) \\ Y_{21}(0) \end{pmatrix} = \begin{pmatrix} -\alpha \\ \alpha - \frac{1}{2} \\ \alpha \end{pmatrix}.$$

Order ϵ terms in the outer expansion must be determined by the corresponding coefficients in the differential system. Thus,

$$\begin{cases} \dfrac{dX_1}{dt} = -2Y_{20}Y_{21}, \\[2mm] \dfrac{dY_{10}}{dt} = 2Y_{10}Y_{11} - 2X_0X_1, \\[2mm] \dfrac{dY_{20}}{dt} = -2Y_{11} + 2Y_{21} \end{cases}$$

requires us to take

$$Y_{21} = Y_{11} + \frac{1}{2}\frac{dY_{20}}{dt} = Y_{11} + \frac{1}{2}\frac{1}{(1+t)^2},$$

$$Y_{11} = -X_1 - \frac{1}{2}\left(\frac{1}{1+t}\right)$$

and X_1 must satisfy the linear equation

$$\frac{dX_1}{dt} = -\frac{2X_1}{1+t} - \frac{1}{(1+t)^2} + \frac{1}{(1+t)^3}.$$

Note that X_1 will match the corresponding term in the inner expansion if we pick

$$X_1(0) = -\alpha.$$

Thus, we uniquely obtain

$$X_1(t) = \frac{1}{(1+t)^2}[-\alpha - t + \ln(1+t)].$$

[Without matching, or its equivalent, we would be hard-pressed to generate the $O(\epsilon)$ approximate solution for $t > 0$.] Continuing in this way, we could also obtain all terms in both the inner and outer expansions. A simpler procedure, however, is to directly seek a uniformly valid composite expansion.

Exercise

Determine a composite expansion for a bounded solution of the preceding conditionally stable nonlinear system in the form

$$\begin{cases} x(t,\epsilon) = X(t,\epsilon) + \epsilon\xi(\tau,\epsilon), \quad x(0,\epsilon) = 1, \\[2mm] y_1(t,\epsilon) = Y_1(t,\epsilon) + \eta_1(\tau,\epsilon), \quad y_1(0,\epsilon) \sim \sum_{j\geq 0} y_{1j}^{\circ}\epsilon^j, \ y_{10}^{\circ} < 1, \ \text{given}, \\[2mm] y_2(t,\epsilon) = Y_2(t,\epsilon) + \eta_2(\tau,\epsilon), \quad y_2(0,\epsilon) \ \text{to be determined}, \end{cases}$$

where the boundary layer correction terms ξ, η_1, and η_2 all decay exponentially to zero as $\tau = t/\epsilon \to \infty$.

A more geometric understanding of the stable initial manifold (which we have approximated asymptotically) can be obtained by representing v_2 in terms of v_1. This phase-plane approach is motivated by the realization that the initial value problems for the v_{1j}'s are stable, whereas those for the v_{2j}'s are not. Recall that the lowest-order term in the inner expansion satisfied

$$\frac{dv_{10}}{d\tau} = v_{10}^2 - 1 \quad \text{and} \quad \frac{dv_{20}}{d\tau} = -2v_{10} + 2v_{20}.$$

If we obtain v_{20} from integrating the linear equation

$$\frac{dv_{20}}{dv_{10}} = \frac{2(v_{20} - v_{10})}{v_{10}^2 - 1},$$

we obtain

$$\left(\frac{v_{10}+1}{v_{10}-1}\right) v_{20} = c_0 - \ln(v_{10}-1)^2 + \frac{2}{v_{10}-1}.$$

In order for v_{20} to remain defined as $v_{10} \to -1$ (i.e., $\tau \to \infty$), we must pick $c_0 = 1 + \ln 4$ to obtain

$$v_{20}(v_{10}) = 1 + \frac{2(v_{10}-1)}{v_{10}+1} \ln\left(1 - \frac{(v_{10}+1)}{v_{10}-1}\right).$$

Note that this implies that $v_{20} \to -1$ as $\tau \to \infty$ and that the limiting stable initial manifold must satisfy

$$y_{20}^{\circ} = 1 + \left(\frac{y_{10}^{\circ}-1}{y_{10}^{\circ}+1}\right) \ln\left(\frac{2}{y_{10}^{\circ}-1}\right)^2.$$

as we previously determined.

For later terms, we can also replace the independent variable τ by v_{10} and integrate the resulting linear systems. Thus, the first-order correction terms will satisfy

$$
\begin{cases}
(v_{10}^2 - 1)\dfrac{du_1}{dv_{10}} = -v_{20}^2, \quad u_1(y_{10}^\circ) = 0, \\[3mm]
(v_{10}^2 - 1)\dfrac{dv_{11}}{dv_{10}} = 2v_{10}v_{11} - 2u_1, \quad v_{11}(y_{10}^\circ) = y_{11}^\circ, \\[3mm]
(v_{10}^2 - 1)\dfrac{dv_{21}}{dv_{10}} = 2v_{21} - 2v_{11}, \quad \text{with } v_{21}(y_{10}^\circ) = y_{21}^\circ \text{ to be specified.}
\end{cases}
$$

Since $v_{20} \to -1$ as $v_{10} \to -1$, we can define

$$
u_1(v_{10}) = -\int_{y_{10}^\circ}^{v_{10}} \frac{v_{20}^2(r)}{r^2 - 1}\, dr.
$$

Likewise,

$$
v_{11}(v_{10}) = \left(\frac{v_{10}^2 - 1}{(y_{10}^\circ)^2 - 1}\right) y_{11}^\circ - 2(v_{10}^2 - 1)\int_{y_{10}^\circ}^{v_{10}} \frac{u_1(r)}{(r^2 - 1)^2}\, dr
$$

will also be well defined as $v_{10} \to -1$. Integrating the equation for v_{21}, we obtain

$$
\left(\frac{v_{10} + 1}{v_{10} - 1}\right) v_{21} = -2\int^{v_{10}} \frac{v_{11}(r)}{(r - 1)^2}\, dr.
$$

Taking limits as $v_{10} \to -1$, we select

$$
v_{21}(v_{10}) = \frac{2(v_{10} - 1)}{(v_{10} + 1)}\int_{v_{10}}^{-1} \frac{v_{11}(r)}{(r - 1)^2}\, dr.
$$

Note that this requires us to use the initial value

$$
y_{21}^\circ = \frac{2(y_{10}^\circ - 1)}{y_{10}^\circ + 1}\int_{y_{10}^\circ}^{-1} \left(\frac{r + 1}{r - 1}\right)\left[\frac{y_{11}^\circ}{(y_{10}^\circ)^2 - 1}\right.
$$

$$
\left. + 2\int_{y_{10}^\circ}^{r} \frac{1}{(s^2 - 1)^2}\left(\int_{y_{10}^\circ}^{s} \frac{v_{20}^2(t)}{t^2 - 1}\, dt\right) ds\right] dr,
$$

thereby specifying the stable initial manifold to $O(\epsilon)$ terms.

J. Singular Problems

We shall call singular perturbation problems *singular* when the reduced problem has an infinity of solutions. Such problems arise in a number of applications, for example, in chemical kinetics where conservation of mass imposes interrelations between variables. We shall consider initial value problems for a system of n equations

$$\epsilon \frac{dy}{dt} = g(y, t, \epsilon)$$

when the limiting outer problem

$$g(Y_0, t, 0) = 0$$

has an $(n - k)$-dimensional *equilibrium* manifold of solutions which might be described through a k-vector equation

$$\rho(Y_0, t) = 0.$$

Vasil'eva and Butuzov (1980) referred to such problems as singular perturbation problems "in the critical case," while Flaherty and O'Malley (1980) called them "singular singular perturbation" problems. While the vector y is, in general, fast because $\dot{y} = O(1/\epsilon)$, the singularity of the Jacobian g_y implies that some combinations of y coordinates might be slow, i.e., they can have bounded derivatives. Our analysis will try to decouple such fast and slow dynamics.

For simplicity, let us *assume* that the equilibrium manifold can be parameterized by $n - k$ components Y_{20} of the vector Y_0, so that we can replace $\rho = 0$ by the more explicit description

$$Y_{10} = \psi(Y_{20}, t)$$

of the equilibrium manifold which we obtain by solving k of the n equations $g(Y_0, t, 0) = 0$. By reordering, if necessary, let us rewrite our original system as

$$\epsilon \frac{dy_i}{dt} = g_i(y_1, y_2, t, \epsilon) \qquad \text{for } i = 1, 2$$

where $\partial g_1 / \partial y_1$ is a nonsingular matrix on the $(n - k)$-dimensional equilibrium manifold. Since we will then have

$$g_{i0} \equiv g_i(\psi(Y_{20}, t), Y_{20}, t, 0) = 0, \quad \text{for both } i = 1 \text{ and } 2,$$

differentiation with respect to Y_{20} implies that

$$\frac{\partial g_{i0}}{\partial y_1} \frac{d\psi}{dY_{20}} + \frac{\partial g_{i0}}{\partial y_2} = 0.$$

Thus, we have both

$$\frac{\partial \psi}{\partial Y_{20}} = -\left(\frac{\partial g_{10}}{\partial y_1}\right)^{-1}\frac{\partial g_{10}}{\partial y_2}$$

and

$$\frac{\partial g_{20}}{\partial y_2} - \frac{\partial g_{20}}{\partial y_1}\left(\frac{\partial g_{10}}{\partial y_1}\right)^{-1}\frac{\partial g_{10}}{\partial y_2} = 0.$$

Further, the $O(\epsilon)$ terms in an outer expansion will likewise imply that

$$\frac{dY_{i0}}{dt} = \frac{\partial g_{i0}}{\partial y_1}Y_{11} + \frac{\partial g_{i0}}{\partial y_2}Y_{21} + \frac{\partial g_{i0}}{\partial \epsilon}, \qquad i = 1, 2.$$

Solving the first equation for Y_{11} and substituting into the second yields

$$\frac{dY_{20}}{dt} = \frac{\partial g_{20}}{\partial y_1}\left(\frac{\partial g_{10}}{\partial y_1}\right)^{-1}\left(\frac{dY_{10}}{dt} - \frac{\partial g_{10}}{\partial y_2}Y_{21} - \frac{\partial g_{10}}{\partial \epsilon}\right) + \frac{\partial g_{20}}{\partial y_2}Y_{21} + \frac{\partial g_{20}}{\partial \epsilon}.$$

Note that our preceding identity implies that the coefficient of Y_{21} is trivial. Further, replacing dY_{10}/dt by

$$\frac{\partial \psi}{\partial Y_{20}}\frac{dY_{20}}{dt} + \frac{\partial \psi}{\partial t}$$

yields the nonlinear system

$$\left[I + \frac{\partial g_{20}}{\partial y_1}\left(\frac{\partial g_{10}}{\partial y_1}\right)^{-2}\frac{\partial g_{10}}{\partial y_2}\right]\frac{dY_{20}}{dt}$$

$$= \frac{\partial g_{20}}{\partial y_1}\left(\frac{\partial g_{10}}{\partial y_1}\right)^{-1}\left(\frac{\partial \psi}{\partial t} - \frac{\partial g_{10}}{\partial \epsilon}\right) + \frac{\partial g_{20}}{\partial \epsilon}.$$

We shall match the outer limit Y_{20} with the corresponding inner limit to obtain the initial condition $Y_{20}(0)$. Moreover, we shall *assume* that the resulting initial value problem for Y_{20} has a unique solution on $0 \leq t \leq 1$. We note that Gu (1987) shows that the system is nonsingular if the nullspace of the Jacobian g_y has a full set of eigenvectors. (The nonsingular coefficient matrix is the Schur complement of $\partial g_{10}/\partial y_1$.)

The limiting inner problem is given by

$$\frac{dz_{i0}}{d\tau} = g_i(z_{10}, z_{20}, 0, 0), \qquad z_{i0}(0) = y_i(0), \quad i = 1, 2, \ \tau \geq 0.$$

Because g_z is singular [with an $(n-k)$-dimensional nullspace], it is natural to expect this autonomous system to have a *dynamic manifold* of $n-k$ conserved scalar quantities, say

$$\sigma(z_{10}, z_{20}) = \sigma(y_1(0), y_2(0))$$

for all $\tau \geq 0$. {The terminology equilibrium and dynamic manifold is borrowed from the control literature [cf. Kokotovic et al. (1986)], where singular problems were studied in power system modeling and in robotics.} We will assume further that it is possible to parameterize the dynamic manifold by z_{10}, i.e., that we can express the constants of motion in the form

$$z_{20} = \pi(z_{10}).$$

(The implicit function theorem, of course, guarantees this possibility when $\partial\sigma/\partial z_{20}$ is nonsingular.) To specifically determine π is generally not easy. We note, however, that our differential system for $dz_{20}/d\tau$ and the chain rule imply that any such π satisfies the initial value problem for the partial differential equation

$$g_2(z_{10}, \pi, 0, 0) = \frac{\partial\pi}{\partial z_{10}} g_1(z_{10}, \pi, 0, 0), \quad y_2(0) = \pi(y_1(0))$$

or, more generally,

$$\frac{\partial\sigma}{\partial z_{10}} g_1 + \frac{\partial\sigma}{\partial z_{20}} g_2 = 0.$$

We shall determine the function π for a number of specific examples. When we can do so, our limiting inner problem is reduced to solving the kth-order stability problem

$$\frac{dz_{10}}{d\tau} = g_1(z_{10}, \pi(z_{10}), 0, 0), \quad z_{10}(0) = y_1(0)$$

on $\tau \geq 0$. A sufficient condition to guarantee the existence of $z_{10}(\tau)$ for all $\tau \geq 0$ is the strict stability of the matrix $\partial g_1/\partial z_1 + (\partial g_1/\partial z_2)(\partial\pi/\partial z_1)$ along the trajectory i.e., stability of the matrix $\partial g_1/\partial y_1$ in an appropriate domain [cf. Gu (1987)]. Knowing $z_{10}(\infty)$ through integration, we also specify $z_{20}(\infty)$, so matching provides the initial value

$$Y_{20}(0) = \pi(z_{10}(\infty))$$

needed to specify the limiting outer solution. Higher-order terms in these expansions could also be obtained in a straightforward manner under related natural hypotheses.

Clearly, the critical need is to determine the $(n - k)$-dimensional constant of motion $\sigma(z_{10}, z_{20})$ for the limiting inner problem. When $\partial\sigma/\partial z_2$ is nonsingular, we can replace the (presumably) fast variable y_2 with the slow variable $\sigma(y_1, y_2) = \sigma(y_1(0), y_2(0))$. Because

$$\frac{d\sigma}{dt} = \frac{\partial\sigma}{\partial y_1}\frac{dy_1}{dt} + \frac{\partial\sigma}{\partial y_2}\frac{dy_2}{dt} = \frac{1}{\epsilon}\left(\frac{\partial\sigma}{\partial y_1}g_1(y_1, y_2, t, \epsilon) + \frac{\partial\sigma}{\partial y_2}g_2(y_1, y_2, t, \epsilon)\right)$$

will be bounded (since σ is slowly varying), our system becomes transformed to a fast/slow system of the form

$$\begin{cases} \epsilon\dfrac{dy_1}{dt} = h_1(y_1, \sigma, t, \epsilon), \\[2mm] \dfrac{d\sigma}{dt} = h_2(y_1, \sigma, t, \epsilon). \end{cases}$$

The classical Tikhonov–Levinson theory of Section 2D can be applied to this system provided $\partial h_1/\partial y_1$ is stable in the appropriate domain. Thus, we seek to transform singular to regular-singular perturbation problems. Further, if the function ρ defining the equilibrium manifold has a nonsingular Jacobian $\partial\rho/\partial y_1$ (in the appropriate domain), we will also be able to replace y_1 by the fast variable $\rho(y_1, y_2, t)$ (which becomes asymptotically trivial in the outer region), thereby describing our problem in terms of a "purely fast" vector ρ and a "purely slow" vector σ (to lowest order in ϵ). This, of course, is what Riccati transformations previously achieved for linear problems.

Exercises

1. Rewrite the problem

$$\epsilon\dot{y} = \begin{pmatrix} 1-2\epsilon & 2-2\epsilon \\ -1+\epsilon & -2+\epsilon \end{pmatrix} y$$

in terms of a purely fast variable $\rho = y_1+2y_2$ and a purely slow variable $\sigma = y_1 + y_2$. Note that this transformation of coordinates is invertible.

2. Show that the preceding analysis does not apply to the example

$$\epsilon\dot{y} = \begin{pmatrix} 0 & \epsilon \\ -1 & 0 \end{pmatrix} y$$

whose solution is unbounded for $t > 0$ unless $y_1(0) = O(\sqrt{\epsilon})$ and $y_2(0) = O(1)$.

3. Solve the initial value problem for the nonlinear equation

$$\epsilon\dot{y} = \epsilon y - y^3$$

on $t \geq 0$. Determine, in particular, the leading term of the outer expansion and an appropriate stretched variable for the initial layer.

A First Example

For the two-dimensional nonlinear system

$$\begin{cases} \epsilon\dfrac{dy_1}{dt} = y_2 + \epsilon m(y_1, y_2, t, \epsilon), \\[2mm] \epsilon\dfrac{dy_2}{dt} = -y_1 y_2, \end{cases}$$

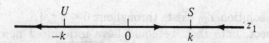

Fig. 2.11. Stability of the rest points $z_1 = \pm k$.

the reduced system implies that the limiting outer solution Y_0 satisfies

$$Y_{20} = 0$$

but leaves Y_{10} free. The $O(\epsilon)$ terms in the outer expansion then imply that

$$\begin{cases} \dfrac{dY_{10}}{dt} = Y_{21} + m(Y_{10}, 0, t, 0), \\[2mm] 0 = -Y_{10}Y_{21}, \end{cases}$$

so Y_{10} satisfies

$$Y_{10}\frac{dY_{10}}{dt} = Y_{10}m(Y_{10}, 0, t, 0).$$

Taking $Y_{10} = \pm\sqrt{Y_{10}^2}$, we have a nonlinear initial value problem for Y_{10}^2. To determine the initial value $Y_{10}(0)$, we employ the limiting inner problem

$$\begin{cases} \dfrac{dz_1}{d\tau} = z_2, & z_1(0) = y_1(0), \\[2mm] \dfrac{dz_2}{d\tau} = -z_1 z_2, & z_2(0) = y_2(0). \end{cases}$$

This system has the constant of motion

$$\sigma(z_1, z_2) = \tfrac{1}{2}z_1^2 + z_2 = \tfrac{1}{2}y_1^2(0) + y_2(0)$$

since $d\sigma/d\tau \equiv 0$. This allows us to eliminate z_2 in terms of z_1^2, leaving the stability problem

$$\frac{dz_1}{d\tau} = -\frac{1}{2}(z_1^2 - k^2), \qquad z_1(0) = y_1(0) \quad \text{for } k = \sqrt{y_1^2(0) + 2y_2(0)}.$$

In order to have a real rest point, we will take

$$y_1^2(0) + 2y_2(0) > 0.$$

Noting the signs of $dz_1/d\tau$ shows where z_1 increases and decreases. See Figure 2.11. Thus, we will reach the stable rest point $z_1(\infty) = k$ as $\tau \to \infty$ provided $y_1(0) > -k$. (The case $k = 0$ is treated analogously.)

Since $Y_{10}(0) = k$ is now specified, the outer solution will be determined by the differential equation

$$\frac{d}{dt}Y_{10}^2 = 2\sqrt{Y_{10}^2}\,m(\sqrt{Y_{10}^2},0,t,0),$$

presuming the solution Y_{10} exists throughout $0 \le t \le 1$.

In retrospect, note that we could have introduced new coordinates $\rho = y_2$ and $\sigma = y_2 + \frac{1}{2}y_1^2$ to obtain $y_1 = \pm\sqrt{2(\sigma - \rho)}\ (= \pm\sqrt{y_1^2})$. Then, our problem would be transformed into an initial value problem for

$$\begin{cases} \dfrac{d\sigma}{dt} = \pm\sqrt{2(\sigma - \rho)}\,m(\pm\sqrt{2(\sigma - \rho)},\rho,t,\epsilon), \\[2mm] \epsilon\dfrac{d\rho}{dt} = \mp\sqrt{2(\sigma - \rho)}\rho. \end{cases}$$

In order for ρ to remain bounded, we must pick the upper sign. Then, ρ will have the trivial outer limit, and the outer limit for σ will be determined through the reduced problem

$$\frac{d\Sigma_0}{dt} = \sqrt{2\Sigma_0}\,m(\sqrt{2\Sigma_0},0,t,0), \quad \Sigma_0(0) = \sigma(0)$$

as long as Σ_0 exists. This, of course, agrees with the preceding.

Exercise

Determine the limiting solution to the system

$$\begin{cases} \epsilon\dfrac{dy_1}{dt} = 2(y_1^2 + y_1)y_2, \\[2mm] \epsilon\dfrac{dy_2}{dt} = y_1^2 + y_1 + \epsilon m(y_1,y_2,t,\epsilon) \end{cases}$$

for various initial values. Note that two different solutions of the reduced problem can be utilized.

Further Examples

1. Consider the system

$$\begin{cases} \epsilon\dfrac{dy_1}{dt} = -y_1\alpha(y_1) + \epsilon m_1(y_1,y_2,t,\epsilon), \\[2mm] \epsilon\dfrac{dy_2}{dt} = y_1\beta(y_1,t) + \epsilon m_2(y_1,y_2,t,\epsilon), \end{cases}$$

where $\alpha > 0$ for all y_1's. The reduced problem implies that

$$Y_{10}(t) = 0,$$

but we must use $O(\epsilon)$ terms in the outer solution to specify Y_{20}. After eliminating Y_{11}, this procedure yields

$$\frac{dY_{20}}{dt} = \frac{1}{\alpha(0)} m_1(0, Y_{20}, t, 0)\beta(0, t) + m_2(0, Y_{20}, t, 0).$$

We will determine the initial value for Y_{20} through matching. The limiting inner problem

$$\frac{dz_{10}}{d\tau} = -z_{10}\alpha(z_{10}),$$

$$\frac{dz_{20}}{d\tau} = z_{10}\beta(z_{10}, 0)$$

implies the phase plane equation

$$\frac{dz_{20}}{dz_{10}} = -\frac{\beta(z_{10}, 0)}{\alpha(z_{10})}.$$

Since $z_{20}(y_1(0)) = y_2(0)$, we explicitly obtain the following representation of the dynamic manifold:

$$z_{20} - y_2(0) = -\int\limits_{y_1(0)}^{z_{10}} \frac{\beta(r, 0)}{\alpha(r)} dr.$$

Since $z_{10} = 0$ is the only possible rest point for our system, we must have

$$Y_{20}(0) = z_{20}(\infty) = y_2(0) + \int\limits_{0}^{y_1(0)} \frac{\beta(r, 0)}{\alpha(r)} dr.$$

Solving the resulting initial value problem for $Y_{20}(t)$ then provides the limiting solution for $t > 0$ as long as it continues to exist.

2. Finally, to show how analogous the solution methods are for singular and conditionally stable problems, consider the example:

$$\begin{cases} \epsilon\dot{u} = u^2 - e^u + \epsilon M_1(u, v, w, t, \epsilon), \\ \epsilon\dot{v} = e^u v + w + \epsilon M_2(u, v, w, t, \epsilon), \\ \epsilon\dot{w} = 2(1 - u)e^u v - 2u^2 v - 2uw + \epsilon M_3(u, v, w, t, \epsilon). \end{cases}$$

The limiting solution must satisfy the reduced system which is equivalent to the two conditions

$$U_0^2 - e^{U_0} = 0$$

and

$$e^{U_0} V_0 + W_0 = 0.$$

Plotting U_0^2 and e^{U_0}, we find only one crossing at, say, $U_0(t) = a < 0$. See Figure 2.12. Moreover, this implies that

$$W_0(t) = -a^2 V_0(t),$$

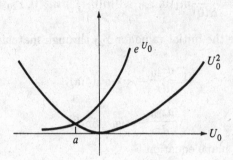

Fig. 2.12. Crossing of e^{U_0} and U_0^2 at $a \approx -0.7$.

but it leaves V_0 unspecified. After some manipulation, the $O(\epsilon)$ terms provide a differential equation for V_0, viz.,

$$\dot{V}_0 = \frac{1}{a(2-a)}[2V_0 M_1(a, V_0, -a^2V_0, t, 0) + 2aM_2(a, V_0, -a^2V_0, t, 0)$$

$$+ M_3(a, V_0, -a^2V_0, t, 0)].$$

The limiting inner problem is given by

$$\begin{cases} \dfrac{d\alpha}{d\tau} = \alpha^2 - e^\alpha, \\[2mm] \dfrac{d\beta}{d\tau} = e^\alpha\beta + \gamma, \\[2mm] \dfrac{d\gamma}{d\tau} = 2(1-\alpha)e^\alpha\beta - 2\alpha^2\beta - 2\alpha\gamma. \end{cases}$$

Note that $\alpha(\tau)$ will decay monotonically to the unique rest point a. Further, the system has $\gamma + 2\alpha\beta$ as a constant of motion, so we will have

$$\gamma(\tau) = w(0) + 2u(0)v(0) - 2\alpha(\tau)\beta(\tau).$$

This leaves us with the conditionally stable system

$$\frac{d\alpha}{d\tau} = \alpha^2 - e^\alpha,$$

$$\frac{d\beta}{d\tau} = (e^\alpha - 2\alpha)\beta + w(0) + 2u(0)v(0).$$

Using the monotonic relationship between α and τ, we change variables and integrate the linear equation

$$\frac{d\beta}{d\alpha} = \frac{(e^\alpha - 2\alpha)\beta + w(0) + 2u(0)v(0)}{\alpha^2 - e^\alpha}$$

to obtain

$$\beta(e^\alpha - \alpha^2) = -[w(0) + 2u(0)v(0)][\alpha - u(0)] + v(0)[e^{u(0)} - u^2(0)].$$

Since $\alpha \to a$ as $\tau \to \infty$, boundedness of β requires $w(0)$ to satisfy

$$w(0) = \frac{v(0)}{a - u(0)}[u^2(0) + e^{u(0)} - 2au(0)].$$

This defines the limiting stable initial manifold and implies that

$$\beta(\tau) = \frac{v(0)(u^2(0) - e^{u(0)})}{a - u(0)} \left(\frac{\alpha(\tau) - a}{e^{\alpha(\tau)} - \alpha^2(\tau)} \right).$$

Thus, as $\tau \to \infty$, we obtain the needed initial value

$$V_0(0) = \beta(\infty) = \frac{v(0)[u^2(0) - e^{u(0)}]}{[a - u(0)](e^a - 2a)}.$$

For more details, see Gu (1987).

As an alternative solution technique, note that we could replace the variable w with the slow variable

$$\sigma = w + 2uv.$$

Since $d\sigma/dt = M_3 + 2M_1v + 2uM_2$, the limiting solution Σ_0 will satisfy the reduced problem

$$\frac{d\Sigma_0}{dt} = M_{30} + \frac{2\Sigma_0}{a(2 - a)}M_{10} + 2aM_{20}, \qquad \Sigma_0(0) = w(0) + 2u(0)v(0)$$

[with the M_{j0}'s evaluated at $(a, \Sigma_0/a(2 - a), -a\Sigma_0/(2 - a))$].

Chapter 3

Singularly Perturbed Boundary Value Problems

A. Second-Order Linear Equations (without Turning Points)

Consider the two-point problem

$$\epsilon y'' + a(x)y' + b(x)y = f(x)$$

on $0 \le x \le 1$ where $a(x) > 0$ and with the boundary values $y(0)$ and $y(1)$ prescribed. We shall suppose that a, b, and f are arbitrarily smooth, and we shall prove that the asymptotic solution will exist, be unique, and have the form

$$y(x, \epsilon) = Y(x, \epsilon) + \xi(x/\epsilon, \epsilon)$$

where the outer expansion $Y(x, \epsilon)$ has a power series expansion

$$Y(x, \epsilon) \sim \sum_{j=0}^{\infty} Y_j(x)\epsilon^j$$

and the initial layer correction $\xi(\tau, \epsilon)$ has an expansion

$$\xi(\tau, \epsilon) \sim \sum_{j=0}^{\infty} \xi_j(\tau)\epsilon^j$$

such that each ξ_j (and its derivatives) will tend to zero as the stretched variable $\tau = x/\epsilon$ tends to infinity. We shall base our proof on the existence of both a smooth asymptotic solution of the differential equation and of a (linearly independent) asymptotic solution which features an initial layer of nonuniform convergence near $x = 0$. Recognizing that the asymptotic solution is an additive composite function of the slow "time" x and the fast time x/ϵ generalizes to multitime expansions in many asymptotic contexts [cf. Nayfeh (1973) and, especially, Kevorkian and Cole (1981)].

Our ansatz allows us to formally obtain the unique solution directly by a straightforward expansion procedure. Since $\xi \to 0$ as $\tau \to \infty$, the outer solution $Y(x, \epsilon)$ must provide the asymptotic solution for $x > 0$. Thus, it must be a smooth solution of the differential equation

$$\epsilon Y'' + a(x)Y' + b(x)Y = f(x)$$

which asymptotically satisfies the terminal condition

$$Y(1, \epsilon) = y(1).$$

Equating coefficients termwise then determines the terms of the outer expansion successively. Thus, we must have

$$a(x)Y_0' + b(x)Y_0 = f(x), \qquad Y_0(1) = y(1),$$

and

$$a(x)Y_j' + b(x)Y_j = -Y_{j-1}''(x), \qquad Y_j(1) = 0,$$

for each $j > 0$. Specifically, we obtain

$$Y_0(x) = e^{\int_x^1 [b(s)/a(s)]ds} \left[y(1) - \int_x^1 e^{\int_x^r [b(s)/a(s)]ds} \frac{f(r)}{a(r)}dr \right]$$

and

$$Y_j(x) = \int_x^1 e^{\int_x^r [b(s)/a(s)]ds} \frac{Y_{j-1}''(r)}{a(r)}dr, \quad j = 1, 2, \ldots.$$

We note that the limiting outer solution $Y_0(x)$ satisfies the reduced problem obtained by solving the limiting differential equation and the terminal condition (i.e., the initial condition is ignored). Moreover, the outer expansion (which is asymptotically valid for $0 < x \leq 1$) follows termwise without any knowledge of the boundary layer behavior near $x = 0$. Because we cannot expect $Y(0, \epsilon)$ to equal $y(0)$, a nontrivial initial layer correction $\xi(\tau, \epsilon)$ must be anticipated.

Because the outer solution Y and the sum $Y + \xi$ should both satisfy

$$\epsilon \frac{d^2Y}{dx^2} + a(x)\frac{dY}{dx} + b(x)Y = f(x) = \epsilon \left(\frac{d^2Y}{dx^2} + \frac{1}{\epsilon^2}\frac{d^2\xi}{d\tau^2} \right)$$
$$+ a(x)\left(\frac{dY}{dx} + \frac{1}{\epsilon}\frac{d\xi}{d\tau} \right) + b(x)(Y + \xi),$$

the initial layer correction ξ must satisfy the (rescaled) homogeneous equation

$$\frac{d^2\xi}{d\tau^2} + a(\epsilon\tau)\frac{d\xi}{d\tau} + \epsilon b(\epsilon\tau)\xi = 0$$

and the initial condition

$$\xi(0, \epsilon) = y(0) - Y(0, \epsilon)$$

as well as the decay requirement that ξ and its derivatives tend to zero as $\tau \to \infty$. Expanding

$$\xi(\tau, \epsilon) \sim \sum_{j=0}^{\infty} \xi_j(\tau) \epsilon^j$$

in powers of ϵ, we must first have

$$\frac{d^2 \xi_0}{d\tau^2} + a(0) \frac{d\xi_0}{d\tau} = 0$$

whereas higher-order terms ξ_j will satisfy nonhomogeneous equations of the form

$$\frac{d^2 \xi_j}{d\tau^2} + a(0) \frac{d\xi_j}{d\tau} = \beta_{j-1}(\tau),$$

where β_{j-1} is successively known and, by induction, decays exponentially to zero. Integrating from infinity, then, we will have

$$\frac{d\xi_0}{d\tau} + a(0) \xi_0 = 0$$

and

$$\frac{d\xi_j}{d\tau} + a(0) \xi_j = - \int_{\tau}^{\infty} \beta_{j-1}(r) dr.$$

Thus,

$$\xi_0(\tau) = e^{-a(0)\tau} [y(0) - Y_0(0)],$$

whereas

$$\xi_j(\tau) = -e^{-a(0)\tau} Y_j(0) - \int_0^{\tau} e^{-a(0)(\tau-s)} \int_s^{\infty} \beta_{j-1}(r) dr ds$$

for each $j > 0$. In summary, then, we have a unique solution $y(x, \epsilon) = Y_0(x) + \xi_0(\tau) + O(\epsilon)$, i.e.,

$$y(x, \epsilon) = \left(e^{\int_x^1 [b(s)/a(s)] ds} y(1) - \int_x^1 e^{\int_x^r [b(s)/a(s)] ds} \frac{f(r)}{a(r)} dr \right)$$

$$+ e^{-a(0)x/\epsilon} \left(y(0) - e^{\int_0^1 [b(s)/a(s)] ds} y(1) \right.$$

$$\left. + \int_0^1 e^{\int_0^r [b(s)/a(s)] ds} \frac{f(r)}{a(r)} dr \right) + O(\epsilon).$$

For small ϵ, then, we have Figure 3.1.

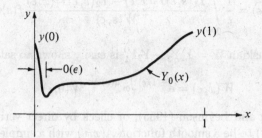

Fig. 3.1. Typical solution with an initial layer.

If, instead, $a(x) < 0$ held throughout the interval, the solution would have a terminal layer of nonuniform convergence of $O(\epsilon)$ thickness and the asymptotic solution would then have the form

$$y(x,\epsilon) = Y(x,\epsilon) + \eta(\sigma,\epsilon),$$

where $\eta \to 0$ as $\sigma = (1-x)/\epsilon \to \infty$. [This follows by introducing $s = 1 - x$ and determining $y(1-s,\epsilon)$ via the preceding.] Here, the outer limit $Y(x,0)$ for $x < 1$ is determined by the limiting differential equation and the initial condition.

If y were a vector and $a(x)$ a nonsingular diagonal matrix $a(x) = \mathrm{diag}[a_{11}(x),\cdots,a_{nn}(x)]$, we would expect the components $y_i(x)$ of y to have boundary layers at one endpoint, according to the sign of the corresponding diagonal entry $a_{ii}(x)$. Such a result even generalizes to diagonally dominant matrices $a(x)$ [cf. Kreiss et al. (1986)].

Much different behavior could, however, result (in the scalar case) if $a(x)$ had a zero at a "turning point" within $(0,1)$. For example, if $a(x)$ changed from negative to positive at, say, $x = \frac{1}{2}$, we might expect (based on the preceding) that the limiting solution would satisfy

$$\begin{cases} a(x)Y_{L0}' + b(x)Y_{L0} = f(x), & Y_{L0}(0) = y(0) \text{ on } 0 \le x < \frac{1}{2}, \\ a(x)Y_{R0}' + b(x)Y_{R0} = f(x), & Y_{R0}(1) = y(1) \text{ on } \frac{1}{2} < x \le 1. \end{cases}$$

Even though these equations will have singular points at $x = \frac{1}{2}$, with suitable boundedness and stability assumptions, we might still naturally seek a transition or shock layer of nonuniform convergence near $\frac{1}{2}$. (See Section 3E for more definite results.)

We shall prove our basic result for the scalar case with $a(x) > 0$ using variation of parameters. If Y_1 and Y_2 are linearly independent solutions of the homogeneous equation $\epsilon y'' + a(x)y' + b(x)y = 0$, we can obtain a particular solution of the nonhomogeneous equation in the form

$$y_p(x, \epsilon) = \frac{1}{\epsilon} \int\limits_0^x \frac{Y_1(s,\epsilon)Y_2(x,\epsilon) - Y_2(s,\epsilon)Y_1(x,\epsilon)}{W(s,\epsilon)} f(s)ds,$$

where the Wronskian $W = Y_1 Y_2' - Y_2 Y_1'$ is easily shown to satisfy

$$W(x, \epsilon) = e^{-(1/\epsilon) \int_0^x a(r)dr} W(0, \epsilon)$$

[cf. Coddington and Levinson (1955), or check by direct verification]. We will take $Y_1(x, \epsilon)$ to be a smooth function $A(x, \epsilon)$ with a simple power series expansion

$$A(x, \epsilon) \sim \sum_{j=0}^\infty A_j(x)\epsilon^j \qquad \text{such that} \ \ A(0, \epsilon) = 1.$$

It, then, satisfies $\epsilon A'' + a(x)A' + b(x)A = 0$ termwise, as well as the initial condition. Thus, we will have

$$a(x)A_0' + b(x)A_0 = 0, \qquad A_0(0) = 1$$

and

$$a(x)A_j' + b(x)A_j = -A_{j-1}'', A_j(0) = 0, \qquad j = 1, 2, \ldots,$$

so

$$A_0(x) = e^{-\int_0^x [b(s)/a(s)]ds}$$

and

$$A_j(x) = -\int\limits_0^x e^{-\int_s^x [b(r)/a(r)]dr} \frac{A_{j-1}''(s)}{a(s)}ds, \qquad j > 0.$$

Note that $A'(0, \epsilon)$ is uniquely determined by the Borel–Ritt theorem [cf. Wasow (1965)] through the formal expansion $A'(0, \epsilon) \sim \sum_{j=0}^\infty A_j'(0)\epsilon^j$. Thus, our preceding theory for singularly perturbed initial value problems guarantees the existence of a unique smooth solution $Y_1(x, \epsilon) = A(x, \epsilon)$ with the asymptotic expansion generated.

We will determine the linearly independent solution $Y_2(x, \epsilon)$ in the convenient WKB form [cf. Olver (1974)]

$$Y_2(x, \epsilon) = B(x, \epsilon)e^{-(1/\epsilon) \int_0^x a(s)ds}.$$

Since $Y_2' = [B' - (1/\epsilon)a(x)B]e^{-(1/\epsilon) \int_0^x a(s)ds}$ and Y_2'' is analogously expressed, the differential equation for Y_2 implies that B must satisfy the singularly perturbed equation

$$\epsilon B'' - a(x)B' - [a'(x) - b(x)]B = 0.$$

As with A, we are guaranteed to have a unique smooth solution $B(x, \epsilon)$ such that $B(0, \epsilon) = 1$ where B has a power series expansion

$$B(x,\epsilon) \sim \sum_{j=0}^{\infty} B_j(x)\epsilon^j.$$

Thus, the positivity of $a(x)$ shows that Y_2 will have boundary layer behavior (i.e., nonuniform convergence as $\epsilon \to 0$) in a $O(\epsilon)$ neighborhood of $x = 0$. The rapid decay of Y_2 implies that it could be reexpanded as a function of the fast variable $\tau = x/\epsilon$ to obtain an asymptotically equivalent expansion

$$Y_2(x,\epsilon) = \eta(\tau,\epsilon) \sim \sum_{j=0}^{\infty} \eta_j(\tau)\epsilon^j,$$

where the η_j's all tend exponentially to zero as $\tau \to \infty$. For only moderately small values of ϵ, however, one would expect the WKB form to provide a better numerical approximation.

Using these solutions $Y_i(x,\epsilon)$, the resulting Wronskian has the initial value $W(0,\epsilon) = Y_2'(0,\epsilon) - Y_1'(0,\epsilon) = -a(0)/\epsilon + B'(0,\epsilon) - A'(0,\epsilon) \equiv -(1/\epsilon)a(0)k(\epsilon)$, where $k(\epsilon) = 1 + O(\epsilon)$. Thus, the particular solution is given by

$$y_p(x,\epsilon) = -\frac{1}{a(0)k(\epsilon)}\int_0^x \Big[A(s,\epsilon)B(x,\epsilon)e^{-(1/\epsilon)\int_0^x a(r)dr}$$

$$- B(s,\epsilon)e^{-(1/\epsilon)\int_0^s a(r)dr}A(x,\epsilon)\Big]e^{(1/\epsilon)\int_0^s a(r)dr}f(s)ds,$$

or

$$y_p(x,\epsilon) = \frac{A(x,\epsilon)}{a(0)k(\epsilon)}\int_0^x B(s,\epsilon)f(s)ds$$

$$- \frac{B(x,\epsilon)}{a(0)k(\epsilon)}\int_0^x A(s,\epsilon)e^{-(1/\epsilon)\int_s^x a(r)dr}\, f(s)ds.$$

The first integral has an asymptotic expansion as a power series in ϵ with smooth coefficients. The asymptotic behavior of the second integral follows from integration by parts, i.e.,

$$-\int_0^x A(s,\epsilon)f(s)e^{-(1/\epsilon)\int_s^x a(r)dr}ds$$

$$= \int_0^x \left(\frac{\epsilon}{a(s)}A(s,\epsilon)f(s)\right)\frac{d}{ds}\left(e^{-(1/\epsilon)\int_s^x a(r)dr}\right)ds$$

$$= \frac{\epsilon}{a(x)}A(x,\epsilon)f(x) - \frac{\epsilon}{a(0)}A(0,\epsilon)f(0)e^{-(1/\epsilon)\int_0^x a(r)dr}$$

$$- \epsilon\int_0^x \frac{d}{ds}\left(A(s,\epsilon)\frac{f(s)}{a(s)}\right)e^{-(1/\epsilon)\int_s^x a(r)dr}ds.$$

Since the last integral has the same form as the original, except for the ϵ factor, the procedure can be repeated indefinitely to show that y_p has an asymptotic representation

$$y_p(x,\epsilon) = C(x,\epsilon) + \epsilon D(x,\epsilon)e^{-(1/\epsilon)\int_0^x a(r)dr},$$

where C and D are smooth functions with asymptotic power series expansions. Any solution of the differential equation differs from y_p by a solution of the homogeneous equation, so if the two-point problem is solvable, the solution must be of the form

$$y(x,\epsilon) = A(x,\epsilon)\ell_1(\epsilon) + B(x,\epsilon)e^{-(1/\epsilon)\int_0^x a(s)ds}\,\ell_2(\epsilon)$$

$$+C(x,\epsilon) + \epsilon D(x,\epsilon)e^{-(1/\epsilon)\int_0^x a(r)dr}$$

for some coefficients $\ell_i(\epsilon)$. Because $A(0,\epsilon) = B(0,\epsilon) = 1$ and $y_p(0,\epsilon) = 0$, the boundary conditions imply that $\ell_1(\epsilon)$ and $\ell_2(\epsilon)$ must satisfy the linear equations

$$y(0) = \ell_1(\epsilon) + \ell_2(\epsilon)$$

and

$$y(1) = A(1,\epsilon)\ell_1(\epsilon) + C(1,\epsilon) + [B(1,\epsilon)\ell_2(\epsilon) + \epsilon D(1,\epsilon)]e^{-(1/\epsilon)\int_0^1 a(s)ds}.$$

This system is, however, uniquely solvable for ϵ sufficiently small. We will have $\ell_2(\epsilon) = y(0) - \ell_1(\epsilon)$ and the remaining equation for $\ell_1(\epsilon)$ has the asymptotic solution

$$\ell_1(\epsilon) \sim \frac{y(1) - C(1,\epsilon)}{A(1,\epsilon)}$$

since $e^{-(1/\epsilon)\int_0^1 a(s)ds}$ is asymptotically negligible, whereas

$$A(1,\epsilon) = e^{-\int_0^1 [b(s)/a(s)]ds} + O(\epsilon) \neq 0$$

for ϵ small. Because these coefficients $\ell_1(\epsilon)$ and $\ell_2(\epsilon)$ have bounded asymptotic power series expansions in ϵ, it follows that the unique solution y of the two-point problem is given by

$$y(x,\epsilon) = Y(x,\epsilon) + E(x,\epsilon)e^{-(1/\epsilon)\int_0^x a(s)ds},$$

where $Y = A\ell_1 + C$ and $E = B\ell_2 + \epsilon D$ have asymptotic expansions in powers of ϵ. Reexpanding the second term as a function of $\tau = x/\epsilon$ shows that y will have the anticipated asymptotic form

$$y(x,\epsilon) = Y(x,\epsilon) + \xi(\tau,\epsilon).$$

This proof, then, relies simply on the corresponding theory for singularly perturbed initial value problems.

An alternative direct proof can be obtained by setting

$$y(x, \epsilon) = \sum_{j=0}^{N} [Y_j(x) + \xi_j(x/\epsilon)]\epsilon^j + \epsilon^{N+1} R(x, \epsilon)$$

and showing existence and uniqueness for the solution of the resulting two-point boundary value problem for the remainder R through use of the equivalent Volterra–Fredholm integral equation [cf. Cochran (1962, 1968) and Smith (1975)].

Exercises

1. Determine the asymptotic form of the solution $y(x)$ for the two-point problem

$$\begin{cases} \epsilon y'' + a(x)y' + b(x)y = f(x), & 0 \le x \le 1, \\ \text{with } y'(0) \text{ and } y(1) \text{ given} \end{cases}$$

when $a(x) > 0$. Note that the function $y(x)$ will converge uniformly throughout $0 \le x \le 1$ as $\epsilon \to 0$ [see O'Malley (1967)].

2. Consider the two-point problem

$$\begin{cases} \epsilon y'' + \mu a(x)y' + b(x)y = f(x), & 0 \le x \le 1, \\ \text{with } y(0) \text{ and } y(1) \text{ prescribed.} \end{cases}$$

Show that the limiting solution within $(0, 1)$ will be $Y_0(x) = f(x)/b(x)$ when $b(x) < 0 < a(x)$ and either (i) $\epsilon/\mu^2 \to 0$ as $\mu \to 0$ or (ii) $\mu^2/\epsilon \to 0$ as $\epsilon \to 0$ [see O'Malley (1967)].

3. Integrate the two-point problem

$$\begin{cases} \epsilon \ddot{x} + (t - \tfrac{1}{2})\dot{x} = 0, & 0 \le t \le 1, \\ x(0), \ x(1) \text{ prescribed} \end{cases}$$

and verify that the unique solution is given by

$$x(t, \epsilon) = x(0) + A(t, \epsilon)[x(1) - x(0)],$$

where

$$A(t, \epsilon) = \frac{\int_0^t e^{-(s-1/2)^2/2\epsilon} ds}{\int_0^1 e^{-(s-1/2)^2/2\epsilon} ds}.$$

Show that the solution can be described in terms of a shock layer at $t = \tfrac{1}{2}$, the stretched variable $\kappa = (t - \tfrac{1}{2})/\sqrt{\epsilon}$, and limits as $\kappa \to \pm\infty$.

4. Determine the constant limiting solution for the two-point problem

$$\begin{cases} \epsilon \ddot{x} - (t - \tfrac{1}{2})\dot{x} = 0, & 0 \le t \le 1, \\ x(0), \ x(1) \text{ prescribed} \end{cases}$$

within $(0, 1)$. *Hint*: One way of proceeding is to show that

$$x(t, \epsilon) = \left(\frac{\int_t^1 e^{(s-1/2)^2/2\epsilon} ds}{\int_0^1 e^{(s-1/2)^2/2\epsilon} ds} \right) x(0) + \left(\frac{\int_0^t e^{(s-1/2)^2/2\epsilon} ds}{\int_0^1 e^{(s-1/2)^2/2\epsilon} ds} \right) x(1)$$

and to asymptotically evaluate this representation as $\epsilon \to 0$ [cf. Erdélyi (1956)]. *Note*: This example relates to an exit time problem for randomly perturbed dynamical systems [cf. Schuss (1980) and Korolyuk (1990)].

B. Linear Scalar Equations of Higher Order

Following Wasow (1941, 1944), let us consider the scalar equation

$$\epsilon^{\ell-k} L(y) + K(y) = f(x)$$

where

$$L(y) = y^{(\ell)} + \sum_{j=1}^{\ell} \alpha_j(x) y^{(\ell-j)}$$

and

$$K(y) = \beta_0(x) y^{(k)} + \sum_{j=1}^{k} \beta_j(x) y^{(k-j)}$$

on the interval $0 \le x \le 1$ where

$$\beta_0(x) \ne 0$$

and the coefficients α_j, β_j, and f are arbitrarily smooth. We shall take

$$\ell > k \ge 0$$

and ask that this ℓth-order singularly perturbed equation be subject to ℓ separated boundary conditions of the form

$$A_i y(0) = \gamma_i, \qquad i = 1, 2, \dots, r;$$

$$A_i y(1) = \gamma_i, \qquad i = r+1, r+2, \dots, r+s = \ell$$

with

$$A_i y = y^{(\lambda_i)} + \sum_{j=0}^{\lambda_i - 1} a_{ij} y^{(j)}$$

for integers λ_i ordered so that

$$\ell > \lambda_1 > \lambda_2 > \cdots > \lambda_r \geq 0$$

and

$$\ell > \lambda_{r+1} > \lambda_{r+2} > \cdots > \lambda_{r+s} \geq 0.$$

Many generalizations of this problem occur in applications, including eigen-value problems, problems with periodic coefficients, with periodic or multi-point boundary conditions, and problems with several singular perturbation parameters in the differential equation or boundary conditions. More abstract operator equations, where the order of the problem is again lowered when $\epsilon = 0$, are considered, e.g., by Vishik and Lyusternik (1960), and such singularly perturbed boundary value problems for partial differential equations are considered by Eckhaus (1979), Zauderer (1983), Carrier and Pearson (1976), and Lions (1973), among others. The essential ideas here already occur in seeking asymptotic solutions when the α_j's, β_j's, and a_{ij}'s for $j > 0$ are all zero. Readers may then simplify details, if they wish, by restricting attention to the less cumbersome problem

$$\begin{cases} \epsilon^{\ell-k} y^{(\ell)} + \beta_0(x) y^{(k)} = f(x), \\ y^{(\lambda_i)}(0) = \gamma_i, \qquad i = 1, \ldots, r, \lambda_i \downarrow, \\ y^{(\lambda_j)}(1) = \gamma_j, \qquad j = r+1, \ldots, r+s = \ell, \lambda_j \downarrow. \end{cases}$$

In the constant coefficient case, a full set of ℓ linearly independent solutions of the homogeneous differential equation can be simply constructed. Then, the associated characteristic polynomial will have $\ell - k$ asymptotically large roots, which tend to a determination of $(1/\epsilon)(-\beta_0)^{1/(\ell-k)}$ as $\epsilon \to 0$, in addition to k bounded roots. Linearly independent asymptotic solutions corresponding to each large stable (or unstable) root can be normalized to display boundary layer behavior near $x = 0$ (or $x = 1$) and k asymptotic solutions can be obtained which are regular perturbations of any set of k linearly independent solutions of the homogeneous reduced equation $K(y_0) = 0$. For $\ell - k$ even, there can, of course, sometimes be a pair of linearly independent rapidly varying oscillatory solutions corresponding to large, purely imaginary roots $(1/\epsilon)(-\beta_0)^{1/(\ell-k)}$.

With variable coefficients, the form of the ℓ linearly independent asymptotic solutions of the homogeneous problem is analogous to that for the constant coefficient problem because each root $[-\beta_0(x)]^{1/(\ell-k)}$ is constant in phase for all x in $[0, 1]$. These solutions then have the form

$$Y_j(x, \epsilon) = G_j(x, \epsilon) e^{(1/\epsilon) \int^x (-\beta_0(s))^{1/(\ell-k)} ds} \qquad \text{for } j = 1, 2, \ldots, \ell - k$$

and

$$Y_j(x, \epsilon) = G_j(x, \epsilon) \qquad \text{for } j = \ell - k + 1, \ldots, \ell,$$

where the G_j's have asymptotic power series expansions in ϵ. The general solution of the nonhomogeneous equation can be obtained by variation of parameters. Whether or not the given boundary value problem has a unique solution as $\epsilon \to 0$ depends, of course, on whether or not a resulting $\ell \times \ell$ determinant is asymptotically nonzero.

Let us *assume* that p (q) of the roots

$$[-\beta_0(x)]^{1/(\ell-k)}$$

have positive (negative) real parts and, to avoid complication, that $p + q = \ell - k$. [The exceptional case that $p + q = \ell - k - 2$ is treated in the works by Wasow (1944) and Handelman et al. (1968).] Let us further *assume* that

(a) $q \leq r$ and $p \leq s$ [so at least q (p) initial (terminal) conditions are prescribed],

(b)   the reduced problem

$$\begin{cases} K(y) = f(x), & 0 \leq x \leq 1, \\ A_i y(0) = \gamma_i, & i = q+1, \ldots, r, \\ A_i y(1) = \gamma_i, & i = r+p+1, \ldots, r+s \end{cases}$$

[which is defined, because of (a)] has a unique solution $Y_0(x)$, and

(c) $\lambda_1, \lambda_2, \ldots, \lambda_q$ (the orders of the cancelled initial conditions) are distinct modulo $\ell - k$, as are $\lambda_{r+1}, \lambda_{r+2}, \cdots, \lambda_{r+p}$.

Then, we shall show that the given boundary value problem will have a unique asymptotic solution of the form

$$y(x, \epsilon) = Y(x, \epsilon) + \epsilon^{\lambda_q} \xi(\tau, \epsilon) + \epsilon^{\lambda_{r+p}} \eta(\sigma, \epsilon)$$

where the outer expansion

$$Y(x, \epsilon) \sim \sum_{j \geq 0} Y_j(x) \epsilon^j$$

provides the asymptotic solution within $0 < x < 1$, where the initial layer correction

$$\epsilon^{\lambda_q} \xi(\tau, \epsilon) \sim \sum_{j \geq 0} \xi_j(\tau) \epsilon^{j + \lambda_q}$$

decays exponentially to zero as the stretched variable $\tau = x/\epsilon$ tends to infinity [thereby providing the nonuniform convergence of the solution and/or its derivatives in an $O(\epsilon)$ initial layer], and where the terminal layer correction

$$\epsilon^{\lambda_{r+p}} \eta(\sigma, \epsilon) \sim \sum_{j \geq 0} \eta_j(\sigma) \epsilon^{j + \lambda_{r+p}}$$

decays to zero as $\sigma = (1 - x)/\epsilon \to \infty$. This asymptotic representation can be differentiated repeatedly to yield

$$y^{(j)}(x, \epsilon) = \frac{d^j Y(x, \epsilon)}{dx^j} + \epsilon^{\lambda_q - j} \frac{d^j \xi(\tau, \epsilon)}{d\tau^j} + (-1)^j \epsilon^{\lambda_{r+p} - j} \frac{d^j \eta(\sigma, \epsilon)}{d\sigma^j}$$

for every $j \geq 0$. The representation can be used as an effective ansatz for the formal construction of asymptotic solutions, as the example which follows illustrates. Note that (b) shows that the limiting solution $Y_0(x)$ within $(0,1)$ can be obtained through a *cancellation law* defined by using the reduced (kth-order) differential equation and the k boundary conditions remaining after the first q initial conditions and the first p terminal conditions are cancelled. The corresponding cancellation law is more complicated in the exceptional case that $p + q < \ell - k$.

Example

Consider the simple problem

$$\begin{cases} \epsilon^2 y^{(4)} - y^{(2)} = x^2, \\ y(0) = 0, \quad y'(0) = 1, \quad y'(1) = 2, \quad y''(1) = 3. \end{cases}$$

Since the homogeneous equation has solutions $1, x, e^{-x/\epsilon}$, and $e^{-(1-x)/\epsilon}$ and a particular solution is given by $-\frac{1}{12}x^4 - \epsilon^2 x^2$, we might expect the boundary value problem to have a solution of the form

$$y(x, \epsilon) = Y(x, \epsilon) + \epsilon\xi(\tau, \epsilon) + \epsilon^2\eta(\sigma, \epsilon)$$

featuring endpoint boundary layers. This agrees with our ansatz since here $[-b_0(x)]^{1/(\ell-k)} = 1^{1/2}$ has one positive and one negative value. Thus, the limiting solution $Y_0(x)$ should be expected to satisfy the reduced problem

$$Y_0'' = -x^2, \qquad Y_0(0) = 0, \qquad Y_0'(1) = 2,$$

thereby determining

$$Y_0(x) = -\frac{x^4}{12} + \frac{7}{3}x.$$

Since the boundary conditions $y'(0) = 1$ and $y''(1) = 3$ were cancelled in defining $Y_0(x)$, we must expect y' (and higher derivatives) to converge nonuniformly at $x = 0$ and $y^{(\ell)}$ to converge nonuniformly at $x = 1$ for all $\ell \geq 2$. This accounts for the powers of ϵ used before ξ and η in our suggested representation of the solution.

We naturally seek an outer solution $Y(x, \epsilon)$ which is a smooth solution of the differential equation

$$\epsilon^2 Y^{(4)} - Y^{(2)} = x^2.$$

Setting $Y \sim \sum_{j \geq 0} Y_j \epsilon^j$ implies that

$$Y_0'' = -x^2,$$
$$Y_1'' = 0,$$
$$Y_2'' = Y_0'''' = -2,$$

etc., so integration provides $Y_0(x) = -\frac{1}{12}x^4 + c_0 + d_0 x, Y_1(x) = c_1 + d_1 x,$
$Y_2(x) = -x^2 + c_2 + d_2 x$, etc. Linearity and the choice of $\tau = x/\epsilon$ as the
stretched variable implies that the initial layer correction must satisfy the
homogeneous equation

$$\frac{d^4\xi}{d\tau^4} - \frac{d^2\xi}{d\tau^2} = 0.$$

Integrating twice backward from $\tau = \infty$ therefore implies that $d^2\xi/d\tau^2 - \xi =$
0. Thus, decaying solutions must have the form

$$\xi(\tau, \epsilon) = e^{-\tau} F(\epsilon),$$

where F has a power series expansion in ϵ. Analogously, the decaying terminal layer correction η will be given by

$$\eta(\sigma, \epsilon) = e^{-\sigma} G(\epsilon),$$

for some G. Thus, our ansatz leads us to look for an asymptotic solution in
the more explicit form

$$y(x, \epsilon) = \left(-\frac{x^4}{12} - \epsilon^2 x^2\right) + C(\epsilon) + D(\epsilon)x + \epsilon e^{-x/\epsilon} F(\epsilon) + \epsilon^2 e^{-(1-x)/\epsilon} \ G(\epsilon).$$

The smooth parts of the solution provide the outer solution and the nonsmooth parts the boundary layer corrections. (For this simple equation, we
could also have guessed this by the method of undetermined coefficients.)
The boundary conditions now imply that

$$y(0, \epsilon) \sim C(\epsilon) + \epsilon F(\epsilon) \sim 0,$$

$$y'(0, \epsilon) \sim D(\epsilon) - F(\epsilon) \sim 1,$$

$$y'(1, \epsilon) \sim -\tfrac{1}{3} - 2\epsilon^2 + D(\epsilon) + \epsilon G(\epsilon) \sim 2$$

$$y''(1, \epsilon) \sim -1 - 2\epsilon^2 + G(\epsilon) \sim 3$$

[where only asymptotically exponentially small terms of order $O(e^{-1/\epsilon})$ are
neglected]. This linear system for $C, D, F,$ and G has a unique solution
which provides us the uniform asymptotic solution

$$y(x, \epsilon) = \left(-\frac{x^4}{12} + \frac{7}{3}x\right) + 4\epsilon\left(-x - \frac{1}{3} + \frac{1}{3}e^{-x/\epsilon}\right)$$

$$+ \epsilon^2(-x^2 + 2x + 4 - 4e^{-x/\epsilon} + 4e^{-(1-x)/\epsilon}) + 2\epsilon^3(-x - 1 + e^{-x/\epsilon})$$

$$+ 2\epsilon^4(1 - e^{-x/\epsilon} + e^{-(1-x)/\epsilon}) + O(e^{-1/\epsilon}).$$

Exercise

Show directly that the following boundary value problems have no limiting
solution as $\epsilon \to 0$:

a. $\epsilon^3 y'''' + y' = 0$, $y''(0) = y'(0) = y(0) = 0, y(1) = 1$;

b. $\epsilon y'' - y' = 0$, $y'(0) = 1, y(1) = 0$;

c. $\epsilon^3 y'''' - y' = 0$, $y'''(0) = y(0) = y(1) = 0, y'(1) = 1$;

d. $\epsilon^2 y''' + y' = 0$, $y'(0) = 0, y(0) = y'(1) = 1$;

e. $\epsilon^2 y'''' + y'' = 0$, $y''(0) = y'(0) = y(1) = 0, y'(1) = 1$.

For our general linear singularly perturbed problem, the outer expansion

$$Y(x,\epsilon) \sim \sum_{j=0}^{\infty} Y_j(x)\epsilon^j$$

should satisfy the scalar equation

$$\epsilon^{\ell-k} L(Y) + K(Y) = f(x)$$

as a power series in ϵ. This implies that successive terms will be smooth
solutions of the kth-order differential equations

$$\begin{cases} K(Y_0) = f(x), \\ K(Y_j) = 0, \quad j = 1, 2, \ldots, \ell - k - 1, \\ K(Y_j) = -L(Y_{j-\ell+k}), \quad j \geq \ell - k. \end{cases}$$

Variation of parameters then determines each Y_j, in turn, up to an arbitrary solution Y_j^c of the homogeneous equation $K(Y_j^c) = 0$. More careful
bookkeeping would indeed show that many Y_j's may be trivial because they
are not needed. Further, the complementary solutions have the form

$$Y_j^c = \sum_{m=1}^{k} \tilde{Y}_m(x) e_{jm}$$

where the $\tilde{Y}_m$'s are linearly independent solutions of $K(Y) = 0$ and the
e_{jm}'s are constants. Thus, the outer solution $Y(x,\epsilon)$ has been shown to lie
on a k-dimensional manifold depending smoothly on ϵ.

Linearity further implies that the (scaled) initial layer correction $\xi(\tau,\epsilon)$
should satisfy the homogeneous equation $\epsilon^{\ell-k} L(\xi) + K(\xi) = 0$. Since
$d^j/dx^j = (1/\epsilon^j)(d^j/d\tau^j)$ for every j, we will have

$$\epsilon^{\ell-k}\left(\frac{1}{\epsilon^\ell}\frac{d^\ell\xi}{d\tau^\ell} + \sum_{j=1}^{\ell}\frac{1}{\epsilon^{\ell-j}}\alpha_j(\epsilon\tau)\frac{d^{\ell-j}\xi}{d\tau^{\ell-j}}\right)$$

$$+\left(\frac{1}{\epsilon^k}\beta_0(\epsilon\tau)\frac{d^k\xi}{d\tau^k} + \sum_{j=1}^{k}\frac{1}{\epsilon^{k-j}}\beta_j(\epsilon\tau)\frac{d^{k-j}\xi}{d\tau^{k-j}}\right) = 0.$$

Multiplying by ϵ^k, we then have

$$\frac{d^\ell\xi}{d\tau^\ell} + \beta_0(\epsilon\tau)\frac{d^k\xi}{d\tau^k} + \sum_{j=1}^{\ell}\epsilon^j\alpha_j(\epsilon\tau)\frac{d^{\ell-j}\xi}{d\tau^{\ell-j}} + \sum_{j=1}^{k}\epsilon^j\beta_j(\epsilon\tau)\frac{d^{k-j}\xi}{d\tau^{k-j}} = 0$$

on $\tau \geq 0$. Expanding

$$\xi(\tau,\epsilon) \sim \sum_{j=0}^{\infty}\xi_j(\tau)\epsilon^j$$

as a power series in ϵ shows that the leading term ξ_0 will need to be a decaying solution of the constant coefficient equation

$$\frac{d^\ell\xi_0}{d\tau^\ell} + \beta_0(0)\frac{d^k\xi_0}{d\tau^k} = 0,$$

whereas higher-order terms will be decaying solutions of nonhomogeneous equations

$$\frac{d^\ell\xi_j}{d\tau^\ell} + \beta_0(0)\frac{d^k\xi_j}{d\tau^k} = \gamma_{j-1}(\tau),$$

where γ_{j-1} is a linear combination of (derivatives of) preceding terms $\xi_h, h < j$, with coefficients that are polynomials in τ. Suppose the distinct roots $\mu_j = [-\beta_0(0)]^{1/(\ell-k)}$ are ordered so that $\text{Re}\,\mu_j < 0$ for $j = 1, 2, \ldots, q$. Then, the decaying solution ξ_0 must have the form

$$\xi_0(\tau) = \sum_{j=1}^{q}e^{\mu_j\tau}c_{j0}.$$

By induction, we find that the later decaying terms ξ_j will be a sum of quasipolynomials

$$\xi_j(\tau) = \sum_{m=1}^{q}e^{\mu_m\tau}(c_{mj} + \tau\delta_{m,j-1}(\tau)),$$

where the c_{mj}'s are undetermined constants and each $\delta_{m,j-1}$ is a polynomial in τ of limited degree which is completely specified by preceding terms. Thus, we have found a q-dimensional manifold of possible initial layer corrections, which depends smoothly on the small parameter ϵ.

Analogously, the terminal layer correction

$$\eta(\sigma, \epsilon) \sim \sum_{j=0}^{\infty} \eta_j(\sigma) \epsilon^j$$

will be determined in terms of the p distinct roots

$$\nu_j = [-\beta_0(1)]^{1/(\ell-k)}, \qquad j = 1, 2, \ldots, p,$$

which satisfy $\mathrm{Re}\,\nu_j > 0$. Specifically, we will have

$$\eta_j(\sigma) = \sum_{m=1}^{p} e^{-\nu_m \sigma} (r_{mj} + \sigma \gamma_{m,j-1}(\sigma)),$$

where the r_{mj}'s are constants and each $\gamma_{m,j-1}$ is known successively as a polynomial in σ. Altogether, then, we have generated a k manifold of outer solutions, a q-manifold of solutions which decay rapidly to zero in an $O(\epsilon)$ thick initial layer, and a p-manifold of solutions which decay rapidly in an analogous terminal layer. The ℓ boundary conditions must now be used to uniquely determine the $\ell = k + p + q$ unspecified constants in the successive terms Y_j, ξ_j, and η_j of the asymptotic solution.

Because the terminal layer correction η and its derivatives are asymptotically negligible near $x = 0$, our ansatz implies that we will have

$$y^{(j)}(0) \sim Y^{(j)}(0, \epsilon) + \epsilon^{\lambda_q - j} \frac{d^j \xi(0, \epsilon)}{d\tau^j} \qquad \text{for all } j.$$

This, in turn, implies that the initial conditions have the asymptotic form

$$\gamma_i = A_i y(0, \epsilon) \sim Y^{(\lambda_i)}(0, \epsilon) + \epsilon^{\lambda_q - \lambda_i} \frac{d^{\lambda_i} \xi(0, \epsilon)}{d\tau^{\lambda_i}}$$

$$+ \sum_{j=0}^{\lambda_i - 1} a_{ij} \left(Y^{(j)}(0, \epsilon) + \epsilon^{\lambda_q - j} \frac{d^j \xi(0, \epsilon)}{d\tau^j} \right).$$

Since $\lambda_q > \lambda_i$ when $i > q$, the last $r - q$ initial conditions with $\epsilon = 0$ imply that the limiting solution Y_0 will satisfy the initial conditions

$$A_i Y_0(0) = \gamma_i, \qquad i = q + 1, \ldots, r.$$

Coefficients of higher powers of ϵ likewise imply that later terms Y_k will satisfy nonhomogeneous initial conditions

$$A_i Y_k(0) = \delta_{k-1}$$

where δ_{k-1} is known successively in terms of initial values of earlier terms in the outer expansion and the initial layer correction and their derivatives. Since $\lambda_i > \lambda_q$ when $q > i$, the first q initial conditions, upon rescaling by multiplication by $\epsilon^{\lambda_i - \lambda_q}$, yield

$$\frac{d^{\lambda_i}\xi(0,\epsilon)}{d\tau^{\lambda_i}} \sim \epsilon^{\lambda_i - \lambda_q}[\gamma_i - A_i Y(0,\epsilon)] - \sum_{j=0}^{\lambda_i - 1} \epsilon^{\lambda_i - j} a_{ij} \frac{d^j \xi(0,\epsilon)}{d\tau^j}.$$

When $\epsilon = 0$, we obtain the following q initial conditions for the leading term ξ_0 of the initial layer expansion

$$\begin{cases} \dfrac{d^{\lambda_i}\xi_0(0)}{d\tau^{\lambda_i}} = 0, & i = 1, 2, \ldots, q-1, \\[3mm] \dfrac{d^{\lambda_q}\xi_0(0)}{d\tau^{\lambda_q}} = \gamma_q - A_q Y_0(0). \end{cases}$$

Because $\xi_0(\tau) = \sum_{j=1}^q e^{\mu_j \tau} c_{jo}$, $d^j \xi_0(0)/d\tau^j = \sum_{m=1}^q \mu_m^j c_{mo}$. This provides q linear equations for the q unknowns c_{mo}, viz.,

$$\sum_{m=1}^q \mu_m^{\lambda_i} c_{mo} = \delta_{qi}(\gamma_q - A_q Y_0(0)), \quad i = 1, 2, \ldots, q,$$

using the Kronecker function with $\delta_{mm} = 1$ and $\delta_{mi} = 0$ otherwise. Unique solvability follows if the coefficient matrix

$$(\mu_m^{\lambda_i})$$

is nonsingular. Since these μ_m's are $(\ell - k)$th roots of $-\beta_0(0)$, the corresponding *Vandermonde* determinant [cf. e.g., Bellman (1970)] is nonzero if and only if the integers $\lambda_1, \lambda_2, \ldots, \lambda_q$ are distinct modulo $\ell - k$. When this is so, it follows that $\xi_0(\tau)$ is completely specified by the first q initial conditions and $A_q Y_0(0)$. Analogously, when $\lambda_{r+1}, \lambda_{r+2}, \ldots, \lambda_{r+p}$ are distinct modulo $\ell - k$, $\eta_0(\sigma)$ is completely determined by the first p terminal conditions and $A_p Y_0(1)$. The remaining $r - q$ initial conditions and $s - p$ terminal conditions likewise imply that the limiting outer solution $Y_0(x)$ must satisfy the kth-order two-point boundary value problem

$$\begin{cases} K(Y_0) = f(x), \\[2mm] A_i Y_0(0) = \gamma_i, & i = q+1, \ldots, r, \\[2mm] A_i Y_0(1) = \gamma_i, & i = r+p+1, \ldots, r+s. \end{cases}$$

Using any fundamental set $\tilde{Y}_1, \ldots, \tilde{Y}_k$ of solutions to the homogeneous reduced equation $K(Y) = 0$, it follows that this reduced problem is uniquely solvable if and only if

$$\det \begin{pmatrix} A_m \tilde{Y}_j(0) \\ A_n \tilde{Y}_j(1) \end{pmatrix} \neq 0, \quad \begin{matrix} m = q+1, \ldots, r, \\ n = r+p+1, \ldots, r+s, \\ j = 1, \ldots, k. \end{matrix}$$

Higher-order terms in all these expansions are also uniquely determined, in turn, under the same assumptions. We note that Vishik and Lyusternik (1961) show how similar reasoning can be used to solve analogous boundary value problems for partial differential equations.

C. First-Order Linear Systems

Let us now seek an asymptotic solution of the singularly perturbed homogeneous linear system

$$\begin{cases} \dot{x} = A(t)x + B(t)y, \\ \epsilon \dot{y} = C(t)x + D(t)y \end{cases}$$

on the finite interval $0 \leq t \leq 1$ subject to the boundary conditions

$$M \begin{pmatrix} x(0) \\ y(0) \end{pmatrix} + N \begin{pmatrix} x(1) \\ y(1) \end{pmatrix} = \ell.$$

Such a problem could arise by rewriting certain singularly perturbed boundary value problems for higher-order linear scalar differential equations as such first-order systems. We will, for example, later convert the scalar equation $\epsilon^3 z'''' + z' = 0$ into such a singularly perturbed system with a scalar x and a three-dimensional vector y. More generally, we might consider this problem as representative of many high-dimensional linear systems written as

$$\Omega(\epsilon) \dot{u} = E(t)u,$$

where u is a vector and Ω is a block-diagonal matrix

$$\Omega(\epsilon) = \operatorname{diag}(I_0, \epsilon_1 I_1, \epsilon_1 \epsilon_2 I_2, \ldots, \epsilon_1 \epsilon_2 \cdots \epsilon_p I_p),$$

where all the ϵ_i's are small positive parameters and the I_j's are identity matrices. Thus, ϵ is used to illustrate the dependence of solutions on fast dynamics which might frequently be neglected in cruder lower-order models. More sophisticated modelers could, indeed, effectively involve a sequence of small parameters.

We will take x to be an m vector; y, an n vector; and ℓ, an $(m+n)$ vector. Moreover, we will *assume* that the matrices A, B, C, and D are smooth functions of t and that the matrix D is nonsingular with $k > 0$ eigenvalues strictly in the left half-plane and with its remaining $n - k > 0$ eigenvalues in the right half-plane for *all* t in $0 \leq t \leq 1$. These conditions on D, which provide the system conditional stability, are often referred to as a *hyperbolicity* assumption. Solutions of the corresponding reduced system

$$\begin{cases} \dot{X}_0 = A(t)X_0 + B(t)Y_0, \\ 0 = C(t)X_0 + D(t)Y_0 \end{cases}$$

will be given by

$$Y_0(t) = -D^{-1}(t)C(t)X_0(t)$$

where X_0 satisfies the mth-order system

$$\dot{X}_0 = [A(t) - B(t)D^{-1}(t)C(t)]X_0.$$

We can expect any limiting solution to our two-point problem to satisfy this reduced-order system. When such a limiting solution exists throughout the open interval $(0,1)$, it cannot generally satisfy the prescribed $m + n$ boundary conditions. Which boundary conditions should be used to specify the appropriate X_0 is certainly not obvious.

We shall transform our system so that (i) m slow (and n fast) modes are decoupled and (ii) k fast-decaying and $n - k$ fast-growing modes are also decoupled. This will allow the construction of a fundamental matrix of asymptotic solutions, selected so that m modes are slow, k modes decay rapidly to zero in an initial $[O(\epsilon)$ thick] layer of nonuniform convergence, and the remaining $n - k$ modes are asymptotically significant only in a corresponding terminal layer region. The relevant theory could, indeed, be conveniently expressed in terms of the exponential dichotomy concept [cf., e.g., Coppel (1978), Ascher et al. (1988), and Jeffries and Smith (1989)]. Whether the given boundary value problem has a unique solution or not will depend on the invertibility of the appropriate $(m+n) \times (m+n)$ -dimensional linear system of algebraic equations. The asymptotic result will follow quite simply due to the form of the boundary layer solutions and because any limiting solution within $(0,1)$ will be a linear combination of only the m slow modes.

We will now obtain a fundamental matrix of asymptotic solutions in a manner analogous to the transformation method used in Section 2B to solve initial value problems for the same linear system. To find a purely fast combination, v, of x and y, we set

$$v = y + L(t, \epsilon)x.$$

Differentiation implies that v will satisfy the purely fast n-dimensional system

$$\epsilon\dot{v} = (D + \epsilon LB)v$$

provided L is a solution of the singularly perturbed matrix Riccati equation

$$\epsilon\dot{L} = DL - C - \epsilon L(A - BL).$$

Note that in our earlier discussion of decoupling for initial value problems, D was a stable matrix. Here, D is simply nonsingular. Likewise,

$$u = x + \epsilon H(t, \epsilon)v$$

will satisfy the purely slow system

$$\dot{u} = (A - BL)u$$

provided H is a solution of the singularly perturbed linear matrix (or Liapunov) equation

$$\epsilon \dot{H} = -H(D + \epsilon LB) - B + \epsilon(A - BL)H.$$

Using matrix notation, we observe that

$$\begin{pmatrix} u \\ v \end{pmatrix} = \begin{pmatrix} I & \epsilon H \\ 0 & I \end{pmatrix} \begin{pmatrix} I & 0 \\ L & I \end{pmatrix} \begin{pmatrix} x \\ y \end{pmatrix}$$

has the inverse transformation

$$\begin{pmatrix} x \\ y \end{pmatrix} = P(t, \epsilon) \begin{pmatrix} u \\ v \end{pmatrix}$$

for

$$P(t, \epsilon) = \begin{pmatrix} I & 0 \\ -L & I \end{pmatrix} \begin{pmatrix} I & -\epsilon H \\ 0 & I \end{pmatrix} = \begin{pmatrix} I & -\epsilon H \\ -L & I + \epsilon LH \end{pmatrix}.$$

In terms of the new variables, then, the system

$$\dot{u} = (A - BL)u,$$

$$\epsilon \dot{v} = (D + \epsilon LB)v$$

has been decoupled while the boundary conditions

$$MP(0, \epsilon) \begin{pmatrix} u(0) \\ v(0) \end{pmatrix} + NP(1, \epsilon) \begin{pmatrix} u(1) \\ v(1) \end{pmatrix} = \ell$$

remain coupled.

To obtain a smooth transformation matrix $P(t, \epsilon)$, we let the matrices L and H have asymptotic expansions

$$L(t, \epsilon) \sim \sum_{j=0}^{\infty} L_j(t)\epsilon^j$$

and

$$H(t, \epsilon) \sim \sum_{j=0}^{\infty} H_j(t)\epsilon^j$$

and determine coefficients termwise from the matrix differential equations as before. Thus,

$$DL_0 - C = 0,$$

$$DL_1 = \dot{L}_0 + L_0(A - BL_0),$$

and

$$-H_0 D - B = 0$$

imply that

$$L_0 = D^{-1}C,$$
$$L_1 = D^{-1}(D^{-1}C)^\bullet + D^{-2}CA - D^{-2}CBD^{-1}C,$$

and

$$H_0 = -BD^{-1}.$$

That a solution exists with the formal series we construct termwise can be shown using integral equations methods [cf., e.g., Harris (1973) and our exercises in Section 2B].

To obtain a fundamental matrix for the decoupled system, we first find a smooth fundamental matrix $\mathcal{U}(t, \epsilon)$ for the mth-order slow system using the initial value problem

$$\dot{\mathcal{U}} = (A - BL)\mathcal{U}, \qquad \mathcal{U}(0, \epsilon) = I.$$

Introducing the series expansion

$$\mathcal{U}(t, \epsilon) \sim \sum_{j=0}^{\infty} \mathcal{U}_j(t)\epsilon^j$$

and equating terms successively implies that the $\mathcal{U}_j$'s will satisfy the initial value problems

$$\dot{\mathcal{U}}_0 = (A - BD^{-1}C)\mathcal{U}_0, \qquad \mathcal{U}_0(0) = I$$

and

$$\dot{\mathcal{U}}_j = (A - BD^{-1}C)\mathcal{U}_j - B\left(\sum_{k=0}^{j-1} L_{j-k}\mathcal{U}_k\right), \quad \mathcal{U}_j(0) = 0$$

for each $j > 0$. Thus, $\mathcal{U}_0$ is a fundamental matrix for the mth-order system satisfied by X_0 and $P(t, 0)\binom{\mathcal{U}_0}{0}$ provides an n-dimensional manifold of smooth solutions for the limiting system for $\binom{X_0}{Y_0}$. Note that variation of parameters implies that

$$\mathcal{U}_j(t) = -\int_0^t \mathcal{U}_0(t)\mathcal{U}_0^{-1}(s)B(s)\left(\sum_{k=0}^{j-1} L_{j-k}(s)\mathcal{U}_k(s)\right) ds$$

for each $j > 0$.

Decomposing n space into the k-dimensional stable and $(n - k)$-dimensional unstable eigenspaces of $D(t)$, we can anticipate that the fast system

$$\epsilon\dot{v} = (D + \epsilon LB)v$$

will have k linearly independent fast-decaying vector solutions and $n - k$ linearly independent fast-growing solutions for every t. In order to obtain the k fast-decaying modes near $t = 0$, we introduce the fast time

$$\tau = t/\epsilon$$

and seek a power series solution

$$V(\tau,\epsilon) \sim \sum_{j=0}^{\infty} V_j(\tau)\epsilon^j$$

of the stretched $n \times k$ matrix system

$$\frac{dV}{d\tau} = [D(\epsilon\tau) + \epsilon L(\epsilon\tau,\epsilon)B(\epsilon\tau)]V$$

such that the terms V_j all tend to zero as $\tau \to \infty$. Expanding the system matrix $D + \epsilon LB$ in a power series in ϵ shows that V_0 must satisfy the homogeneous system

$$\frac{dV_0}{d\tau} = D(0)V_0$$

and successive V_j's must satisfy a nonhomogeneous system

$$\frac{dV_j}{d\tau} = D(0)V_j + \gamma_{j-1}(\tau),$$

where γ_{j-1} is a linear combination of preceding V_k's (and their derivatives) with coefficients which are polynomials in τ. Thus, we will take

$$V_0(\tau) = e^{D(0)\tau}V_0(0)$$

and

$$V_j(\tau) = \int_0^{\tau} e^{D(0)(\tau-s)}\gamma_{j-1}(s)ds$$

for each $j > 0$. In particular, note that the k columns of $V(\tau,\epsilon)$ are then multiples of the corresponding columns of $V(0,\epsilon) = V_0(0)$. Recall that the exponential matrix has the form

$$e^{D(0)\tau} = [e^{\lambda_1\tau}p_1(\tau) \ e^{\lambda_2\tau}p_2(\tau) \cdots e^{\lambda_b\tau}p_b(\tau)]C,$$

where the λ_ℓ's are the b distinct eigenvalues of $D(0)$ and $p_\ell(\tau)$ is an $n \times m_\ell$ matrix polynomial in τ spanning the m_ℓ-dimensional (possibly generalized) eigenspace corresponding to λ_ℓ, with $\sum_{\ell=1}^{b} m_\ell = n$, and where C is a non-singular matrix selected so that $e^0 = I$ [cf., e.g., Coddington and Levinson (1955)]. Let us take Re $\lambda_i < 0$ (> 0) for $i \leq c$ ($i > c$). Then, the first $k = \sum_{i=1}^{c} m_i$ columns of $e^{D(0)\tau}$ will decay exponentially to zero and the last $n - k$ columns will grow exponentially as $\tau \to \infty$. To obtain all the decaying solutions, we introduce

$$V_0(0) = S_0 \equiv [p_1(0) \ p_2(0) \ \cdots \ p_c(0)]$$

as an $n \times k$ matrix whose column space spans the stable eigenspace of $D(0)$. Thus, $V(\tau,\epsilon)$ is defined as the solution of the matrix initial value problem

$$\frac{dV}{d\tau} = [D(\epsilon\tau) + \epsilon L(\epsilon\tau, \epsilon)B(\epsilon\tau)]V, \quad V(0, \epsilon) = S_0.$$

In completely analogous fashion, we introduce the stretched variable

$$\sigma = \frac{1-t}{\epsilon}.$$

Solutions of our fast system featuring terminal boundary layers will necessarily satisfy the stretched system

$$\frac{dw}{d\sigma} = -[D(1 - \epsilon\sigma) + \epsilon L(1 - \epsilon\sigma, \epsilon)B(1 - \epsilon\sigma)]w$$

and must decay to zero as $\sigma \to \infty$. Suppose the columns of an $n \times (n - k)$ matrix S_1 span the $(n - k)$-dimensional unstable eigenspace of $D(1)$. Then w will necessarily lie in the column span of the asymptotic solution

$$W(\sigma, \epsilon) \sim \sum_{j=0}^{\infty} W_j(\sigma)\epsilon^j$$

of the matrix initial value problem

$$\frac{dW}{d\sigma} = -(D + \epsilon LB)W, \qquad W(0, \epsilon) = S_1.$$

Putting our solutions together provides the (asymptotic) fundamental matrix

$$\Phi(t, \epsilon) = \begin{pmatrix} \mathcal{U}(t, \epsilon) & 0 & 0 \\ 0 & V(\tau, \epsilon) & W(\sigma, \epsilon) \end{pmatrix}$$

for the decoupled slow/fast system for u and v and, thereby, the fundamental matrix

$$\Omega(t, \epsilon) = P(t, \epsilon)\Phi(t, \epsilon)$$

for the original x–y system. We note how naturally the solution space decomposes into the sum of an m manifold of smooth outer solutions, a k manifold of initial layer corrections, and an $(n - k)$ manifold of terminal layer corrections. Further, linearity implies that any solution of the boundary value problem must have the form

$$\begin{pmatrix} x \\ y \end{pmatrix} = P(t, \epsilon)\Phi(t, \epsilon)q(\epsilon)$$

for some $(m + n)$ vector $q(\epsilon)$. The boundary conditions then imply that q must satisfy the linear algebraic system

$$\Delta(\epsilon)q(\epsilon) \equiv [MP(0, \epsilon)\Phi(0, \epsilon) + NP(1, \epsilon)\Phi(1, \epsilon)]q(\epsilon) = \ell.$$

If $\Delta(\epsilon)$ is nonsingular (for ϵ sufficiently small), we will have a unique solution

$$q(\epsilon) = \Delta^{-1}(\epsilon)\ell.$$

In particular, when $\Delta(0)$ is nonsingular, the solution $\binom{x}{y}$ will be bounded. However, when $\Delta(\epsilon) \sim K\epsilon^j$ for some $j > 0$ and some constant $K \neq 0, q(\epsilon)$ and the solution $\binom{x}{y}$ will be algebraically unbounded as $\epsilon \to 0$.

An asymptotic expansion for $\Delta(\epsilon)$ can be readily obtained. Neglecting only the asymptotically exponentially small terms $V(1/\epsilon, \epsilon)$ and $W(1/\epsilon, \epsilon)$, we will have

$$\Phi(0, \epsilon) \sim \begin{pmatrix} I & 0 & 0 \\ 0 & S_0 & 0 \end{pmatrix} \quad \text{and} \quad \Phi(1, \epsilon) \sim \begin{pmatrix} \mathcal{U}(1, \epsilon) & 0 & 0 \\ 0 & 0 & S_1 \end{pmatrix},$$

so

$$\Delta(\epsilon) \sim M \begin{bmatrix} I & -\epsilon H(0, \epsilon) S_0 & 0 \\ -L(0, \epsilon) & [I + \epsilon L(0, \epsilon) H(0, \epsilon)] S_0 & 0 \end{bmatrix}$$

$$+ N \begin{bmatrix} \mathcal{U}(1, \epsilon) & 0 & -\epsilon H(1, \epsilon) S_1 \\ -L(1, \epsilon) \mathcal{U}(1, \epsilon) & 0 & [I + \epsilon L(1, \epsilon) H(1, \epsilon)] S_1 \end{bmatrix}.$$

If we now partition M and N like P, we will have

$$\Delta(0) = \begin{bmatrix} \delta_{11} & M_{12} S_0 & N_{12} S_1 \\ \delta_{21} & M_{22} S_0 & N_{22} S_1 \end{bmatrix},$$

where $\delta_{k1} = M_{k1} - M_{k2} D^{-1}(0) C(0) + [N_{k1} - N_{k2} D^{-1}(1) C(1)] \mathcal{U}_0(1)$ for $k = 1$ and 2. *Assuming* now that $\Delta(0)$ is nonsingular and writing

$$\Delta^{-1}(\epsilon) \equiv \Gamma(\epsilon) = \begin{bmatrix} \Gamma_1(\epsilon) \\ \Gamma_2(\epsilon) \\ \Gamma_3(\epsilon) \end{bmatrix},$$

we obtain a unique solution of the form

$$\binom{x}{y} = P(t, \epsilon) \Phi(t, \epsilon) \Gamma(\epsilon) \ell = Z(t, \epsilon) + \xi(\tau, \epsilon) + \eta(\sigma, \epsilon)$$

with the outer expansion

$$Z(t, \epsilon) = \begin{bmatrix} I \\ -L(t, \epsilon) \end{bmatrix} \mathcal{U}(t, \epsilon) \Gamma_1(\epsilon) \ell,$$

the exponentially decaying initial layer correction

$$\xi(\tau, \epsilon) = \begin{bmatrix} -\epsilon H(\epsilon\tau, \epsilon) \\ I + \epsilon L(\epsilon\tau, \epsilon) H(\epsilon\tau, \epsilon) \end{bmatrix} V(\tau, \epsilon) \Gamma_2(\epsilon) \ell,$$

and the terminal layer correction

$$\eta(\sigma, \epsilon) = \begin{bmatrix} -\epsilon H(1 - \epsilon\sigma, \epsilon) \\ I + \epsilon L(1 - \epsilon\sigma, \epsilon) H(1 - \epsilon\sigma, \epsilon) \end{bmatrix} W(\sigma, \epsilon) \Gamma_3(\epsilon) \ell.$$

In terms of the slow and fast vectors u and v, note that our limiting solution within $(0, 1)$ is the solution of the reduced problem

$$\begin{cases} \dot{U}_0 = (A - BD^{-1}C)U_0, \quad U_0(0) = \Gamma_1(0)\ell, \\ V_0 = 0. \end{cases}$$

In terms of the original variables, this corresponds to the reduced system

$$\begin{cases} \dot{X}_0 = AX_0 + BY_0, \\ 0 = CX_0 + DY_0 \end{cases}$$

and the m boundary conditions

$$\Gamma_1(0)\left[M\begin{pmatrix} X_0(0) \\ Y_0(0) \end{pmatrix} + N\begin{pmatrix} X_0(1) \\ Y_0(1) \end{pmatrix} - \ell \right] = 0$$

which we could obtain by multiplying the original vector boundary condition through by $\Delta^{-1}(0)$ and "cancelling" the last n components [cf. Harris (1960)].

Example

Wasow's theory for scalar equations (cf. Section 3B) implies that the boundary value problem

$$\begin{cases} \epsilon^3 z'''' + z' = 0, \qquad 0 \le t \le 1, \\ \\ \text{with} \\ \\ z'(0), z(0), z'(1), \text{ and } z(1) \text{ prescribed} \end{cases}$$

has a unique solution of the form

$$z(t, \epsilon) = Z(t, \epsilon) + \epsilon\xi(\tau, \epsilon) + \eta(\sigma, \epsilon),$$

where $\xi \to 0$ as $\tau = t/\epsilon \to \infty$ and $\eta \to 0$ as $\sigma = (1 - t)/\epsilon \to \infty$, due to the nonuniform convergence of (derivatives) of z at the endpoints. Moreover, the limiting outer solution $Z_0(t)$ will satisfy the reduced problem

$$Z_0' = 0, \qquad Z_0(0) = z(0).$$

Regular perturbation arguments show that the outer solution is, indeed, simply

$$Z(t, \epsilon) = z(0).$$

Further, the derivatives of the solution have the form

$$\frac{d^j z}{dt^j}(t, \epsilon) \sim \frac{1}{\epsilon^j}\left(\epsilon\frac{d^j \xi}{d\tau^j} + (-1)^j\frac{d^j \eta}{d\sigma^j} \right) \quad \text{for each } j > 0,$$

i.e., they are generally unbounded in the endpoint boundary layers.

We can rewrite this problem as a first-order singularly perturbed system by setting

$$x = z \quad \text{and} \quad y = \begin{pmatrix} \epsilon^2 z''' \\ \epsilon z'' \\ z' \end{pmatrix}.$$

We will then have

$$\begin{cases} \dot{x} = By, \\[2mm] \epsilon \dot{y} = Dy, \\[2mm] M\begin{pmatrix} x(0) \\ y(0) \end{pmatrix} + N\begin{pmatrix} x(1) \\ y(1) \end{pmatrix} = \ell \end{cases}$$

for

$$B = (0 \quad 0 \quad 1),$$

$$D = \begin{pmatrix} 0 & 0 & -1 \\ 1 & 0 & 0 \\ 0 & 1 & 0 \end{pmatrix}, \quad M = \begin{pmatrix} 1 & 0 & 0 & 0 \\ 0 & 0 & 0 & 1 \\ 0 & 0 & 0 & 0 \\ 0 & 0 & 0 & 0 \end{pmatrix}, \quad N = \begin{pmatrix} 0 & 0 & 0 & 0 \\ 0 & 0 & 0 & 0 \\ 1 & 0 & 0 & 0 \\ 0 & 0 & 0 & 1 \end{pmatrix},$$

and

$$\ell = \begin{pmatrix} z(0) \\ z'(0) \\ z(1) \\ z'(1) \end{pmatrix}.$$

Other transformations to system form are certainly also possible.

Since y is already a purely fast variable, there is no need here to use a Riccati matrix L. Introducing

$$u = x + \epsilon H y,$$

we will obtain $\dot{u} = \dot{x} + \epsilon \dot{H} y + \epsilon H \dot{y} = (B + \epsilon \dot{H} + HD)y$. Thus, u will be purely slow if H satisfies the linear matrix equation $\epsilon \dot{H} + HD + B = 0$. A direct solution of $H_0 D = -B$ yields the constant solution

$$H = (1 \quad 0 \quad 0).$$

Moreover, the resulting slow system $\dot{u} = 0$ will then have the fundamental matrix

$$\mathcal{U}(t, \epsilon) = 1.$$

The transformation to purely slow and fast variables u and y thus has the inverse

$$\begin{pmatrix} x \\ y \end{pmatrix} = P(\epsilon)\begin{pmatrix} u \\ y \end{pmatrix}$$

with

$$P(\epsilon) = \begin{pmatrix} 1 & -\epsilon & 0 & 0 \\ 0 & 1 & 0 & 0 \\ 0 & 0 & 1 & 0 \\ 0 & 0 & 0 & 1 \end{pmatrix}.$$

The system matrix D for y has the stable eigenvalue -1 and the unstable eigenvalues $\omega = e^{\pi i/3}$ and $\bar{\omega}$ with corresponding eigenvectors

$$\begin{pmatrix} 1 \\ -1 \\ 1 \end{pmatrix}, \quad \begin{pmatrix} -\bar{\omega} \\ \omega \\ 1 \end{pmatrix}, \quad \text{and} \quad \begin{pmatrix} -\omega \\ \bar{\omega} \\ 1 \end{pmatrix},$$

respectively. Thus, the initial layer system

$$\frac{dv}{d\tau} = Dv$$

has the decaying solution

$$V(\tau) = \begin{pmatrix} 1 \\ -1 \\ 1 \end{pmatrix} e^{-\tau},$$

whereas the terminal layer system

$$\frac{dw}{d\sigma} = -Dw$$

has the decaying solutions

$$\begin{pmatrix} -\bar{\omega} \\ \omega \\ 1 \end{pmatrix} e^{-\omega\sigma} \quad \text{and} \quad \begin{pmatrix} -\omega \\ \bar{\omega} \\ 1 \end{pmatrix} e^{-\bar{\omega}\sigma}.$$

The u–v system, therefore, has the complex-valued fundamental matrix

$$\Phi(t,\epsilon) = \begin{pmatrix} 1 & 0 & 0 & 0 \\ 0 & e^{-\tau} & -\bar{\omega}e^{-\omega\sigma} & -\omega e^{-\bar{\omega}\sigma} \\ 0 & -e^{-\tau} & \omega e^{-\omega\sigma} & \bar{\omega}e^{-\bar{\omega}\sigma} \\ 0 & e^{-\tau} & e^{-\omega\sigma} & e^{-\bar{\omega}\sigma} \end{pmatrix},$$

providing a one-dimensional slow solution space, a one-dimensional space of fast decaying initial layer solutions, and a two-dimensional fast-growing terminal layer solution space. The original system therefore has the fundamental matrix $P(\epsilon)\Phi(t,\epsilon)$ and a general solution of the form

$$\begin{pmatrix} x(t) \\ y(t) \end{pmatrix} = P(\epsilon)\Phi(t,\epsilon)q(\epsilon) = \begin{bmatrix} 1 & -\epsilon e^{-\tau} & \epsilon\bar{\omega}e^{-\omega\sigma} & \epsilon\omega e^{-\bar{\omega}\sigma} \\ 0 & e^{-\tau} & -\bar{\omega}e^{-\omega\sigma} & -\omega e^{-\bar{\omega}\sigma} \\ 0 & -e^{-\tau} & \omega e^{-\omega\sigma} & \bar{\omega}e^{-\bar{\omega}\sigma} \\ 0 & e^{-\tau} & e^{-\omega\sigma} & e^{-\bar{\omega}\sigma} \end{bmatrix} q(\epsilon)$$

with q arbitrary. To satisfy the boundary conditions, we need

$$\Delta(\epsilon)q(\epsilon) \equiv [MP(\epsilon)\Phi(0,\epsilon) + NP(\epsilon)\Phi(1,\epsilon)]q(\epsilon) = \ell.$$

The solution $q(\epsilon) = \Delta^{-1}(\epsilon)\ell$ will be unique if $\Delta(\epsilon)$ is nonsingular. Neglecting only exponentially asymptotically small terms like $e^{-1/\epsilon}$ and $e^{-\omega/\epsilon}$, we have

$$P(\epsilon)\Phi(0,\epsilon) \sim \begin{bmatrix} 1 & -\epsilon & 0 & 0 \\ 0 & 1 & 0 & 0 \\ 0 & -1 & 0 & 0 \\ 0 & 1 & 0 & 0 \end{bmatrix}$$

and likewise for $P(\epsilon)\Phi(1,\epsilon)$. Thus, $\Delta(\epsilon)$ has the asymptotic power series expansion

$$\Delta(\epsilon) \sim \begin{bmatrix} 1 & -\epsilon & 0 & 0 \\ 0 & 1 & 0 & 0 \\ 1 & 0 & \epsilon\bar\omega & \epsilon\omega \\ 0 & 0 & 1 & 1 \end{bmatrix}.$$

Note that $\Delta(\epsilon)$ is nearly singular since it has an eigenvalue of order ϵ. Solving the linear system for $q(\epsilon)$, we obtain $q_1 \sim z(0) + \epsilon z'(0)$, $q_2 \sim z'(0)$, and $q_3 = \bar q_4 \sim (i/\sqrt{3}\epsilon)[z(1) - z(0)] - (i/\sqrt{3})[z'(0) + \omega z'(1)]$. Thus, the asymptotic solution is given by

$$\begin{pmatrix} x \\ y_1 \\ y_2 \\ y_3 \end{pmatrix} \sim \begin{pmatrix} 1 \\ 0 \\ 0 \\ 0 \end{pmatrix} [z(0) + \epsilon z'(0)] + \begin{pmatrix} -\epsilon \\ 1 \\ -1 \\ 1 \end{pmatrix} e^{-\tau} z'(0)$$

$$+ \frac{2}{\sqrt{3}\epsilon}\, \text{Im}\left[\begin{pmatrix} \epsilon\bar\omega \\ -\bar\omega \\ \omega \\ 1 \end{pmatrix} e^{-\omega\sigma}[z(1) - z(0) - \epsilon(z'(0) + \omega z'(1))] \right].$$

D. An Application in Control Theory

In the so-called *linear regulator problem*, one seeks to solve the initial value problem

$$\dot x = A(t)x + B(t)u, \quad x(0) \text{ prescribed}$$

on an interval $0 \leq t \leq 1$, selecting the optimal control $u(t)$ to minimize the scalar performance index

$$J(u) = \frac{1}{2} \int\limits_0^1 [x'Q(t)x + u'R(t)u]dt,$$

where the matrices Q and R are symmetric and, respectively, positive semidefinite and positive definite for all t. Here, we will take the state x to be an n vector and the control u to be an r vector. Usually, $n \geq r$. The prime denotes transposition.

Conditions for optimality can be obtained for this and more general problems by classical variational arguments [cf. Kwakernaak and Sivan (1972)], through the Hamilton–Jacobi equations using the Hamiltonian

$$h = \tfrac{1}{2}(x'Qx + u'Ru) + p'(Ax + Bu)$$

[cf. Athans and Falb (1966)], or by using the Pontryagin maximum principle [cf. Fleming and Rishel (1975)]. One finds that the optimal control is given by

$$u = -R^{-1}B'p,$$

the control which minimizes h, where the adjoint (or costate) vector p satisfies the terminal value problem

$$\dot{p} = -Qx - A'p, \qquad p(1) = 0.$$

One could try to directly solve the resulting two-point boundary value problem consisting of the Hamiltonian system

$$\begin{pmatrix} \dot{x} \\ \dot{p} \end{pmatrix} = \begin{pmatrix} A & -BR^{-1}B' \\ -Q & -A' \end{pmatrix} \begin{pmatrix} x \\ p \end{pmatrix}$$

and the boundary conditions that $x(0)$ is given and $p(1) = 0$. Alternatively, one could introduce the feedback $p = K(t)x$ or, equivalently,

$$u(t) = -R^{-1}(t)B'(t)K(t)x(t)$$

where the square matrix $K(t)$ is naturally asked to be the symmetric solution of the terminal value problem

$$\dot{K} = -KA - A'K + KBR^{-1}B'K - Q, \quad K(1) = 0.$$

If the solution K of this matrix Riccati equation exists throughout the interval $0 \leq t \leq 1$ (which is actually guaranteed under the definiteness conditions on Q and R when the coefficients are smooth), we need only solve the linear initial value problem

$$\dot{x} = (A - BR^{-1}B'K)x, \qquad x(0) \text{ prescribed.}$$

Thus, our two-point problem corresponding to optimality is replaced by a matrix terminal value problem for K, followed by an initial value problem for x. This is a major accomplishment, analytically and for purposes of numerical computation. Under natural hypotheses, we are guaranteed to obtain a unique positive definite gain matrix $K(t)$ such that the optimal cost will be given by $x'(0)K(0)x(0)$. We note that other related problems could also be considered, including the time-invariant problem on the semi-infinite interval $t \geq 0$.

An often-studied singular perturbation problem results when the state equation is singularly perturbed [cf. O'Malley (1978), Kokotovic (1984), Kokotovic et al. (1986), and Bensoussan (1988)]. Let us, instead, consider singular problems where $R(t) \equiv 0$ [cf. Bell and Jacobson (1975) and Clements and Anderson (1978)]. Then in the performance index, the control usually involves an endpoint impulse, and the resulting state lies on a lower-dimensional "singular arc" within $0 < t < 1$. Let us seek its solution as the limit of solutions to the *cheap control* problem obtained for

$$R = \epsilon^2 S,$$

with a nonsingular S, in the $\epsilon \to 0$ limit. In the performance index, the control u will then be relatively cheap compared to the state x. We note that analogous regularization methods are often used to study ill-posed problems for purposes of proving the existence of solutions or of computing them [cf., e.g., Kreiss and Lorenz (1989)]. The cheap control problem is, however, of independent interest [cf. Lions (1973)] as are similar high gain problems with K or u unbounded [cf. Utkin (1978) and related work describing actuators]. We shall solve the resulting singularly perturbed matrix Riccati equation

$$\epsilon^2 \dot{k} + \epsilon^2(kA + A'k + Q) - kBS^{-1}B'k = 0,$$

subject to $k(1, \epsilon) = 0$, and the resulting singularly perturbed vector problem

$$\epsilon^2 \dot{x} = (-BS^{-1}B'k + \epsilon^2 A)x,$$

with $x(0)$ given, in order to define the optimal control

$$u = -\frac{1}{\epsilon^2}S^{-1}B'kx.$$

Exercise

Consider the scalar problem with the state equation

$$\dot{x} = u,$$

$x(0) = 1$, and cost functional

$$J(u) = \frac{1}{2} \int\limits_0^1 (x^2 + \epsilon^2 u^2)\,dt.$$

a. Solve the two-point problem

$$\begin{cases} \epsilon^2 \dot{x} = -p, & x(0) = 1, \\ \dot{p} = -x, & p(1) = 0 \end{cases}$$

corresponding to the optimal control $u = -p/\epsilon^2$. Show that the control is given by

$$u = -\frac{(1 + e^{-2/\epsilon})^{-1}}{\epsilon} \left(e^{-t/\epsilon} - e^{-1/\epsilon} e^{-(1-t)/\epsilon} \right)$$

and explain why the limiting control

$$\lim_{\epsilon \to 0} u = \lim_{\epsilon \to 0} \left(-\frac{1}{\epsilon} e^{-t/\epsilon} \right)$$

acts like the delta-function impulse $-\delta(t)$, driving the initial state $x(0)$ instantaneously to zero in a cost-free manner. (For any smooth f, note that $(1/\epsilon) \int_0^1 f(t) e^{-t/\epsilon}\,dt \to f(0)$ as $\epsilon \to 0$.)

b. Obtain the corresponding Riccati gain k as the solution of the scalar problem $\epsilon^2 \dot{k} = k^2 - \epsilon^2, k(1) = 0$. [Look for a solution of the form $k = \epsilon + \epsilon \ell(\sigma, \epsilon)$ where $\ell \to 0$ as $\sigma = (1 - t)/\epsilon \to \infty$.] Then, obtain the optimal control $u = -kx/\epsilon^2$ as a function of t and ϵ.

Returning to our original cheap control problem with $R = \epsilon^2 S$, let us seek a smooth outer solution $K(t, \epsilon)$ of the singularly perturbed matrix Riccati equation

$$\epsilon^2 \dot{K} + \epsilon^2 (KA + A'K + Q) - KBS^{-1}B'K = 0$$

as a power series

$$K(t, \epsilon) \sim \sum_{j \geq 0} K_j(t) \epsilon^j.$$

Since the leading term K_0 must satisfy the reduced equation $K_0 BS^{-1}B'K_0 = 0$,

$$B'K_0 = 0,$$

i.e., K_0 is restricted to lie in the nullspace of B'. When $r = n$ and B is invertible, this implies that $K_0 = 0$. More generally, however, we encounter a singular singular perturbation problem where K_0 is not completely specified by the reduced equation. To describe its equilibrium manifold, one needs to proceed further algebraically. [One might also use a matrix pseudoinverse of B' or, more simply, a matrix which projects onto the range of B.] For simplicity here, we will *assume* that

$$B'QB \quad \text{is nonsingular.}$$

Then, we will actually specify K_0 by manipulating the $O(\epsilon^2)$ term

$$\dot{K}_0 + K_0 A + A' K_0 + Q = K_1 B S^{-1} B' K_1$$

in the Riccati equation. Since $K_0 B = 0$, $\dot{K}_0 B = -K_0 \dot{B}$. Thus, postmultiplication by B implies that $-K_0 \dot{B} + K_0 AB + QB = K_1 BS^{-1} B' K_1 B$. Premultiplication by B' then shows that $B' K_1 BS^{-1} B' K_1 B = B'QB$ is positive definite, so $B' K_1 B = \sqrt{S^{1/2} B'QB S^{1/2}}$ and

$$K_1 B = (K_0 B_1 + QB)(B' K_1 B)^{-1} S$$

with $B_1 \equiv AB - \dot{B}$. Thus, $K_1 BS^{-1} B' K_1 = (K_0 B_1 + QB)(B'QB)^{-1}(B'Q + B_1' K_0)$ and K_0 must satisfy the parameter-free Riccati problem

$$\dot{K}_0 + K_0 A_1 + A_1' K_0 - K_0 B_1 (B'QB)^{-1} B_1' K_0 + Q_1 = 0, \quad K_0(1) = 0,$$

where

$$A_1 \equiv A - B_1 (B'QB)^{-1} B'Q$$

and

$$Q_1 \equiv Q[I - B(B'QB)^{-1} B'Q] = P'QP \geq 0$$

for the projection

$$P \equiv I - B(B'QB)^{-1} B'Q.$$

Standard linear regulator theory [cf. Kwakernaak and Sivan (1972)] guarantees the existence of the positive semidefinite solution $K_0(t)$ throughout $0 \leq t \leq 1$ which, by its construction, will also satisfy $B' K_0 = 0$.

Successive K_j's will satisfy linear variational systems of the form

$$\dot{K}_j + K_j [A_1 - B_1 (B'QB)^{-1} B_1' K_0] + [A_1 - B_1 (B'QB)^{-1} B_1' K_0]' K_j = \alpha_{j-1}.$$

They can be successively obtained analogously or by equating appropriate coefficients in the matrix differential equation

$$\dot{K} + KA + A'K + Q = [KB_1 + QB + (KB)^{\cdot} + A'KB]$$

$$\times [B'QB + B'(KB)^{\cdot} + B'KB_1 + B'A'KB]^{-1}$$

$$\times [B_1'K + B'Q + (B'K)^{\cdot} + B'KA].$$

This equation follows, under the assumption that $B'QB$ is positive definite, by solving the original Riccati equation for $B'KB$ and $B'K$ and using back substitution. We note that the outer expansion will not satisfy the terminal condition $K(1,\epsilon) = 0$ since $(1/\epsilon)B'KB$ is positive-definite (at least for ϵ

sufficiently small). Thus, we naturally anticipate nonuniform convergence of the Riccati gain k near $t = 1$ and set

$$k(t, \epsilon) = K(t, \epsilon) + \epsilon \ell(\sigma, \epsilon),$$

asking that the terminal layer correction $\ell \to 0$ as $\sigma = (1 - t)/\epsilon \to \infty$. Then, ℓ must satisfy the matrix Riccati equation

$$\frac{d\ell}{d\sigma} = -\frac{1}{\epsilon}(\ell BS^{-1}B'K + KBS^{-1}B'\ell) - \ell BS^{-1}B'\ell + \epsilon(\ell A + A'\ell)$$

and its leading term will satisfy

$$\frac{d\ell_0}{d\sigma} = -\ell_0 B(1)S^{-1}(1)B'(1)K_1(1)$$
$$- K_1(1)B(1)S^{-1}(1)B'(1)\ell_0 - \ell_0 B(1)S^{-1}(1)B'(1)\ell_0.$$

The initial value

$$B'(1)\ell_0(0) = -B'(1)K_1(1) = -S(1)[B'(1)K_1(1)B(1)]^{-1}B'(1)Q(1)$$

is specified and since $\ell_0 \to 0$ as $\sigma \to \infty$, we obtain (much like solving scalar Riccati equations)

$$\ell_0(\sigma) = -2K_1(1)B(1)S^{-1/2}(1)(I + e^{2C_0\sigma})^{-1}C_0^{-1}S^{-1/2}(1)B'(1)K_1(1),$$

where $C_0 = S^{-1/2}(1)B'(1)K_1(1)B(1)S^{-1/2}(1)$ is positive definite. We observe that this explicit solution procedure defines the dynamic manifold of the fast Riccati solutions. Note that we have obtained the terminal value $K_1(1) = -\ell_0(0)$ which we need to specify the first-order term K_1 in the outer expansion. Likewise, later terms ℓ_j in this boundary layer correction can be shown to satisfy linear systems and to have exponential decay as $\sigma \to \infty$.

Knowing the Riccati gain, we must still solve the singularly perturbed initial value problem

$$\epsilon \dot{x} = \left[-BS^{-1}B'\left(\frac{K}{\epsilon} + \ell\right) + \epsilon A\right]x, \quad x(0) \text{ prescribed}$$

to determine the state trajectory. We can expect the limiting state X_0 to satisfy the reduced equation $BS^{-1}B'K_1X_0 = 0$, implying that

$$B'K_1X_0 = 0,$$

so X_0 must lie in the nullspace of $B'K_1$. We have again encountered a singular singular perturbation problem whenever $\text{rank}(B'K_1) < n$. Because $B'K_1X_0 = 0$ will usually be incompatible with the prescribed initial condition, we generally need an initial layer correction for x. Likewise, the rapidly varying coefficient $-BS^{-1}B'\ell$ will usually require x to have a higher-order terminal layer correction. Thus, we naturally expect the state vector to have the asymptotic form

$$x(t) = X(t, \epsilon) + \xi(\tau, \epsilon) + \epsilon\eta(\sigma, \epsilon)$$

with an outer solution X, an initial layer correction ξ which decays to zero as $\tau = t/\epsilon \to \infty$, and a terminal layer correction $\epsilon\eta$ whose terms decay to zero as $\sigma = (1 - t)/\epsilon \to \infty$. Details involving the construction of such expansions are given in O'Malley and Jameson (1975, 1977). Note that the control law will then imply a corresponding asymptotic expansion

$$u(t) = U(t, \epsilon) + \frac{1}{\epsilon}v(\tau, \epsilon) + w(\sigma, \epsilon)$$

of the optimal control where the outer expansion

$$U(t, \epsilon) = -\frac{1}{\epsilon^2}S^{-1}B'KX$$

is bounded since $B'K_0 = 0$ and $B'K_1X_0 = 0$. The boundary layer corrections $v(\tau, \epsilon) = -(1/\epsilon)S^{-1}B'K\xi$ and $w(\sigma, \epsilon) = -(1/\epsilon)S^{-1}B'(K\eta + \ell X)$ will also be bounded and they will have power series expansions whose terms decay to zero as τ and σ, respectively, tend to infinity. It is important to note that the leading control term $(1/\epsilon)v_0(t/\epsilon)$ acts like a large initial impulse (indeed, a matrix delta-function) which drives the initial state $x(0)$ to $X_0(0)$ in the nullspace of $B'K_1$. For $0 < t < 1$, the limiting control $U_0(t)$ takes on primary significance. The pair (X_0, U_0) indeed coincides with the classical singular arc solution to the original singular control problem [cf. Moylan and Moore (1971)]. Our singular perturbation approach allows us to also solve nearly singular control problems, as we have demonstrated.

When $B'QB$ is singular, one can analyze cases where the initial impulse is of the form $(1/\mu^L)\sum_{j=0}^{L-1} v_j(\tau)\mu^j$ with $\mu = \sqrt[L]{\epsilon}$ and where each $v_j \to 0$ as $\tau = t/\mu \to \infty$. The limiting control then behaves like a linear combination of impulses $\delta(0), \delta'(0), \ldots, \delta^{(L-1)}(0)$, transferring the initial state $x(0)$ to some $X_0(0^+)$ on the singular arc, a manifold of dimension $n - Lr$. For $L > 1$, the corresponding state will also involve an initial impulse. The algebraic theory of such problems [cf. Willems et al. (1986), Saberi and Sannuti (1987), and Geerts (1989)] is very complicated, independent of the singular perturbations approach.

Exercise

Suppose we wish to control the harmonic oscillator

$$\ddot{y} + y = u \quad \text{with } y(0) \text{ and } \dot{y}(0) \text{ prescribed}$$

and we want to pick u to minimize either the cost functional

$$J_1(u) = \frac{1}{2} \int_0^1 [\dot{y}^2(t) + \epsilon^2 u^2(t)]dt$$

or

$$J_2(u) = \frac{1}{2} \int_0^1 [y^2(t) + \epsilon^2 u^2(t)]dt.$$

a. Convert the problem to vector form by introducing $x = \binom{y}{\dot{y}}$. Note that $B'QB > 0$ for J_1, but that $B'QB = 0$ and $B'A'QAB > 0$ for J_2.

b. Carry out the asymptotic solution with cost functional J_1. Note that the limiting control for $t > 0$ is equal to $y(0)$ and that there is an initial delta-function impulse.

c. Do the same for cost functional J_2. Show that, in the $\epsilon \to 0$ limit, the initial impulse is a linear combination of a delta-function and its derivative and that the limiting solution is trivial for $t > 0$.

E. Some Linear Turning Point Problems

(i) We are interested in the scalar equation

$$\epsilon y'' + a(x, \epsilon)y' + b(x, \epsilon)y = 0 \quad \text{on } -1 \le x \le 1$$

when a and b are smooth, $a(x, 0)$ has a simple zero at $x = 0$ where $a_x(0, 0) < 0$, and $a(x, 0) \ne 0$ elsewhere.

We will first examine the simple example

$$\epsilon y'' - xy' + \lambda y = 0$$

where λ is fixed and where the boundary values $y(\pm 1)$ are prescribed. The exact solution can be obtained in terms of the confluent hypergeometric function F [cf. Abramowitz and Stegen (1965) or Whittaker and Watson (1952)], viz.,

$$y(x, \epsilon) = \frac{F\left(-\frac{\lambda}{2}; \frac{1}{2}; \frac{x^2}{2\epsilon}\right)}{F\left(-\frac{\lambda}{2}; \frac{1}{2}; \frac{1}{2\epsilon}\right)} \cdot \frac{y(-1) + y(1)}{2}$$

$$+ \frac{F\left(\frac{1}{2} - \frac{\lambda}{2}; \frac{3}{2}; \frac{x^2}{2\epsilon}\right)}{F\left(\frac{1}{2} - \frac{\lambda}{2}; \frac{3}{2}; \frac{1}{2\epsilon}\right)} \cdot \frac{[y(1) - y(-1)]x}{2}$$

provided the denominators are nonzero. The asymptotic behavior of this solution as $\epsilon \to 0$ follows from the asymptotic expansions of these special functions as their last argument tends to infinity.

Alternatively, note that the transformed variable

$$z(x, \epsilon) = y(x, \epsilon)e^{-x^2/4\epsilon}$$

must satisfy

$$\epsilon z'' + \left(\lambda + \frac{1}{2} - \frac{x^2}{4\epsilon}\right) z = 0.$$

Since the parabolic cylinder (or Weber) functions $D_n(t)$ and $D_{-n-1}(it)$ are linearly independent solutions of $d^2w/dt^2 + (n + \frac{1}{2} - t^2/4)w = 0$ (cf. the same references), z is naturally obtained as a linear combination of $D_\lambda(-x/\sqrt{\epsilon})$ and $D_{-1-\lambda}(ix/\sqrt{\epsilon})$. Moreover, when λ is a non-negative integer, $D_\lambda(z)$ degenerates to $e^{-z^2/4}He_\lambda(z)$ where He_λ is the λth Hermite polynomial. The asymptotic behavior of the solution to our boundary value problem follows from that of $D_n(z)$ as $|z| \to \infty$. Thus, we obtain the asymptotic limit:

$$y(x, \epsilon) \sim \begin{cases} e^{-(1+x)/\epsilon}y(-1) + e^{-(1-x)/\epsilon}y(1) & \text{for } \lambda \neq 0, 1, 2, \ldots \\ \frac{1}{2}x^\lambda[y(1) + (-1)^\lambda y(-1)] + \frac{1}{2}e^{(x^2-1)/2\epsilon}[y(1) - (-1)^\lambda y(-1)] \\ \qquad\qquad \text{for } \lambda = 0, 1, 2, \ldots. \end{cases}$$

The solution, then, has a $O(\epsilon)$ thick boundary layer at both endpoints, as we might have anticipated, since $a(x) = -x$ is positive (negative) at the left (right) endpoint. The limiting solution within $(-1, 1)$ is asymptotically trivial unless $\lambda = 0, 1, 2, \ldots$, when it tends to a specific multiple of x^λ, thereby satisfying the reduced equation $-xY_0' + \lambda Y_0 = 0$. Further, the solution is bounded throughout $[-1, 1]$ and is zero at the turning point $x = 0$.

Exercise

Note that $y = x$ is a solution of $\epsilon y'' - xy' + y = 0$. Use reduction of order to obtain the exact solution of the two-point problem

$$\begin{cases} \epsilon y'' - xy' + y = 0, \\ y(-1) = 2, \quad y(1) = -1 \end{cases}$$

and show that the limiting solution within $(-1, 1)$ is given by $Y_0(x) = -\frac{3}{2}x$. Pictorially, then, we have Figure 3.2.

Other than by introducing special functions, we could use ordinary matching to suggest that the asymptotic solution to our preceding example should have the limiting form

$$y(x) \sim x^\lambda C + e^{-(1+x)/\epsilon}[y(-1) - (-1)^\lambda C] + e^{-(1-x)/\epsilon}[y(1) - C].$$

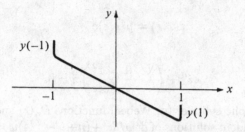

Fig. 3.2. Solution with endpoint layers and nontrivial interior limit.

We are left, however, with no clue concerning how C should be selected. Grasman and Matkowsky (1977) showed that one method of attacking the two-point problem for $\epsilon y'' + a(x, \epsilon)y' + b(x, \epsilon)y = 0$ was to recast it as a variational problem, e.g., to make the functional

$$I(y) = \int_{-1}^{1} L(x, y, y')dx$$

stationary for

$$L = \tfrac{1}{2}[\epsilon(y')^2 - b(x, \epsilon)y^2]e^{(1/\epsilon)\int_0^x a(s, \epsilon)ds}$$

among all smooth functions y satisfying the boundary conditions [cf. also Kevorkian and Cole (1981)]. The Euler–Lagrange equation

$$\frac{d}{dx}\left(\frac{\partial L}{\partial y'}\right) = \frac{\partial L}{\partial y}$$

is a necessary condition to achieve an extremum [cf. Courant and Hilbert (1953)]. This reduces to the given equation $\epsilon y'' + a(x, \epsilon)y' + b(x, \epsilon)y = 0$ since

$$\frac{d}{dx}\left(\frac{\partial L}{\partial y'}\right) = [\epsilon y'' + a(x, \epsilon)y']e^{(1/\epsilon)\int_0^x a(s, \epsilon)ds}.$$

We note that this approach is useful computationally, as well as theoretically [cf. Hemker (1977)]. For our simple example, we use

$$I[y] = \frac{1}{2}\int_{-1}^{1}(\epsilon(y')^2 - \lambda y^2)e^{-(x^2/2\epsilon)}dx.$$

Since the preceding asymptotics imply the limiting behavior of y and y', with

$$y' \sim \lambda x^{\lambda-1}C - \frac{1}{\epsilon}e^{-(1+x)/\epsilon}[y(-1) - (-1)^{\lambda}C] + \frac{1}{\epsilon}e^{-(1-x)/\epsilon}[y(1) - C],$$

we will have

$$I(y) \sim -\frac{1}{2}\lambda C^2 \int_{-1}^{1} x^{2\lambda} e^{-x^2/2\epsilon} dx$$

$$+ \frac{1}{2\epsilon} \left\{ [y(-1) - (-1)^\lambda C]^2 \int_{-1}^{1} e^{-(x+2)^2/2\epsilon} dx \right.$$

$$\left. + [y(1) - C]^2 \int_{-1}^{1} e^{-(x-2)^2/2\epsilon} dx \right\}$$

or

$$I(y) \sim -\lambda C^2 (2\epsilon)^{\lambda+1/2} \int_{0}^{1/\sqrt{2\epsilon}} \kappa^{2\lambda} e^{-\kappa^2} d\kappa$$

$$+ \frac{1}{\sqrt{2\epsilon}} \{ [y(-1) - (-1)^\lambda C]^2 + [y(1) - C]^2 \} \int_{1/\sqrt{2\epsilon}}^{3/\sqrt{2\epsilon}} e^{-\kappa^2} d\kappa.$$

A careful asymptotic analysis (in terms of the incomplete gamma-function or by use of direct methods) shows that the major contribution to I comes from the first integral, unless λ is a non-negative integer [and the resulting coefficient, involving $1/\Gamma(-\lambda)$, becomes trivial]. Usually, then, I takes its extreme value when $C = 0$, so the trivial limit within $(-1,1)$ results. In the exceptional case, we minimize the coefficient of the second term by taking $C = \frac{1}{2}[y(1) + (-1)^\lambda y(-1)]$. {Note that all these exceptional cases can be reduced to the problem with $\lambda = 0$ through differentiation [cf. Kreiss (1981)], so one could also obtain this result by studying that special case.} Then, the limiting solution within $(-1,1)$ is the average of the solutions Y_{L0} and Y_{R0} of the reduced problems

$$xY_0' = \lambda Y_0, \quad Y_{L0}(-1) = y(-1), \quad \text{and} \quad Y_{R0}(1) = y(1).$$

Note how the limiting solution to our example is very sensitive to the specific value of λ. The *resonant* nature of the problem can be further clarified by allowing λ to be a function of ϵ. Sturm–Liouville theory then implies that there are denumerably many eigenvalues $\lambda_k(\epsilon)$ for each $\epsilon > 0$ with a corresponding set of nontrivial eigenfunctions $e_k(x,\epsilon)$ which are zero at both endpoints. Using the Rayleigh quotient would even show that $\lambda_k(\epsilon) \to k$ as $\epsilon \to 0$, for every integer $k > 0$. This explains why nontrivial limiting solutions within $(-1,1)$ occur when λ is one of these eigenvalues [cf. deGroen (1980)]. They also arise when $\lambda(\epsilon)$ is exponentially close to such an integer value [cf. Wasow (1984)].

If we rewrite our general equation in the form

$$\epsilon y'' + x\alpha(x, \epsilon)y' + b(x, \epsilon)y = 0$$

for a negative function $\alpha(x, \epsilon)$, the problem can be transformed to that for a simpler related equation which is asymptotically close to our model problem [cf. Ackerberg and O'Malley (1970) and Wasow (1984), noting the related classical approach of Cherry (1950)]. This approach shows that the limiting solution in $(-1, 1)$ is trivial unless $b(0, 0)/\alpha(0, 0)$ is a nonpositive integer. Higher-order conditions for *resonance* (i.e., a nontrivial limit) are not so simple to write down, but they are equivalent to the existence of a nontrivial power series solution $\sum_{j=0}^{\infty} U_j(x)\epsilon^j$ (i.e., outer expansion) of the equation with coefficients U_j which are smooth throughout $(-1, 1)$ [cf. Matkowsky (1975)]. Any nonsmooth coefficient U_j will eliminate resonance. We note that resonance is sometimes related to certain exit-time problems for stochastic equations [cf. Matkowsky (1980)], but regret to report that the mathematical phenomenon under discussion (despite much attention in the literature) has not yet substantially helped us understand much new physics.

When $\alpha(x, \epsilon)$ is, instead, positive, the limiting solution will tend to

$$\begin{cases} Y_{L0}(x), & -1 \leq x < 0, \\ Y_{R0}(x), & 0 < x \leq 1, \end{cases}$$

where Y_{L0} (Y_{R0}) satisfy the respective reduced problems

$$x\alpha(x, 0)Y_0' + b(x, 0)Y_0 = 0, \quad Y_{L0}(-1) = y(-1), \quad \text{and} \quad Y_{R0}(1) = y(1),$$

provided $b(0, 0)/\alpha(0, 0)$ is not a positive integer. We note that Y_{L0} and/or Y_{R0} may become unbounded as $x \to 0$ and that the limiting solution may be unbounded throughout $[-1, 1]$ in the exceptional case. Consider, for example, the two-point problem for $\alpha \equiv 2$ and $b \equiv 1$.

Exercise

Using parabolic cylinder functions, show that the solution of the two-point problem

$$\begin{cases} \epsilon y'' + xy' + ny = 0, \\ y(\pm 1) \text{ prescribed} \end{cases}$$

will be exponentially large within $(-1, 1)$ when n is a positive integer.

(ii)   Following Kevorkian and Cole (1981), we shall now consider the two-point problem

$$\begin{cases} \epsilon \dfrac{d^2 y}{dx^2} + \sqrt{x} \dfrac{dy}{dx} - y = 0, & 0 \le x \le 1, \\ \\ y(0) = 0, y(1) = e^2 \end{cases}$$

[also see Lagerstrom (1988)]. The endpoint $x = 0$ is a turning point in the sense that it is a singular point of the reduced equation. Since $\sqrt{x} > 0$ for $x \ne 0$, however, we might still anticipate boundary layer behavior at $x = 0$ and convergence elsewhere to an outer solution. Maximum principle and differential inequality arguments [cf. Dorr et al. (1973), Chang and Howes (1984), and Section 3G below] guarantee the existence of a unique bounded solution to the problem.

Introducing an outer expansion

$$Y(x, \epsilon) \sim \sum_{j=0}^{\infty} Y_j(x) \epsilon^j$$

into the differential equation and terminal conditions determines the terms Y_j uniquely and successively as solutions of the first-order terminal value problems

$$\sqrt{x} Y_0' - Y_0 = 0, \qquad Y_0(1) = e^2$$

and, for each $j > 0$,

$$\sqrt{x} Y_j' - Y_j = -Y_{j-1}'', \qquad Y_j(1) = 0.$$

Thus,

$$Y_0(x) = e^{2\sqrt{x}},$$

$$Y_1(x) = e^{2\sqrt{x}} \left(-\frac{1}{2x} + \frac{2}{\sqrt{x}} - \frac{3}{2} \right),$$

and, in general,

$$Y_j(x) = -e^{2\sqrt{x}} \int_1^x \frac{1}{\sqrt{s}} e^{-2\sqrt{s}} Y_{j-1}''(s) ds, \quad j \ge 1,$$

so the Y_j's have the form

$$Y_j(x) = e^{2\sqrt{x}} P_j \left(\frac{1}{\sqrt{x}} \right),$$

where the P_j's are polynomials of increasing order. The singularities at $x = 0$, therefore, become increasingly severe as j increases.

In order to obtain the solution near $x = 0$, we can either seek an inner expansion

$$w(\tau, \epsilon^\beta) \sim \sum_{j=0}^{\infty} w_j(\tau)\epsilon^{j\beta},$$

with β to be determined and

$$\tau = x/\epsilon^\alpha \qquad \text{for some } \alpha > 0,$$

which will match the singular outer expansion as $\tau \to \infty$, or we can directly seek a composite expansion

$$y(x, \epsilon) = Y(x, \epsilon) + u(\tau, \epsilon^\beta)$$

valid throughout $0 \le x \le 1$ with an initial layer correction $u(\tau, \epsilon^\beta) \sim \sum_{j=0}^{\infty} u_j(\tau)\epsilon^{j\beta}$ whose terms decay to zero as $\tau \to \infty$. Since the solution y is bounded, u must cancel the singularities of Y. Linearity implies that both the inner expansion and the expansion for the initial layer correction must satisfy the given equation as functions of τ. Thus,

$$\epsilon^{1-2\alpha}\frac{d^2w}{d\tau^2} + \epsilon^{-\alpha/2}\sqrt{\tau}\frac{dw}{d\tau} - w = 0$$

for all $\tau \ge 0$. Since the middle term has a large coefficient, we must have $1 - 2\alpha = -\alpha/2$, so

$$\alpha = \frac{2}{3},$$

and w (and u) will need to satisfy

$$\frac{d^2w}{d\tau^2} + \sqrt{\tau}\frac{dw}{d\tau} = \epsilon^{1/3}w$$

as a power series in $\epsilon^{1/3}$, i.e., $w(\tau, \epsilon^{1/3}) \sim \sum_{j=0}^{\infty} w_j(\tau)\epsilon^{j/3}$. Since $w(0, \epsilon^{1/3}) = 0$, we obtain the successive initial value problems

$$\frac{d^2w_0}{d\tau^2} + \sqrt{\tau}\frac{dw_0}{d\tau} = 0, \qquad w_0(0) = 0,$$

and, for each $j > 0$,

$$\frac{d^2w_j}{d\tau^2} + \sqrt{\tau}\frac{dw_j}{d\tau} = w_{j-1}, \qquad w_j(0) = 0.$$

Integration implies that

$$w_0(\tau) = c_0 \int_0^\tau e^{-(2/3)s^{3/2}}\,ds$$

and, in general,

$$w_j(\tau) = c_j \int_0^\tau e^{-(2/3)s^{3/2}}\,ds + \int_0^\tau\int_0^s e^{-(2/3)(s^{3/2}-t^{3/2})}w_{j-1}(t)\,dt\,ds$$

for each $j > 0$. (Integrations from $\tau = \infty$ may, sometimes, be preferable.) The constants c_j will have to be carefully determined so the inner and outer expansions match.

First, observe that we will be able to have

$$w_0(\infty) = Y_0(0) = 1$$

if we pick

$$c_0 = \left(\int_0^\infty e^{-(2/3)s^{3/2}} ds \right)^{-1}.$$

(Note that c_0 and w_0 could be expressed in terms of gamma-functions.) Thus, we will have

$$w_0(\tau) = 1 - c_0 \int_\tau^\infty e^{-(2/3)s^{3/2}} ds,$$

so $w_0(\tau) \sim 1$ as $\tau \to \infty$, up to transcendentally small terms, and the first boundary layer correction term is $u_0(\tau) = -c_0 \int_\tau^\infty e^{-(2/3)s^{3/2}} ds$. Repeated integrations by parts imply that

$$\int_\tau^\infty e^{-(2/3)s^{3/2}} ds = [\frac{1}{\tau^{1/2}} - \frac{1}{2}\frac{1}{\tau^2} + \frac{1}{\tau^{7/2}} + O\left(\frac{1}{\tau^5}\right)]e^{-(2/3)\tau^{3/2}}.$$

Substituting this asymptotic behavior into the expression for w_0, then, implies that

$$w_1(\tau) \sim K_1 + \int_\tau^\infty \int_s^\infty e^{-(2/3)s^{3/2}} e^{(2/3)t^{3/2}} dt\,ds$$

$$\sim K_1 + 2\tau^{1/2} - \frac{1}{2\tau} + O\left(\frac{1}{\tau^{5/2}}\right) \quad \text{as } \tau \to \infty$$

for some constant K_1. This, in turn, yields

$$w_2(\tau) \sim \int^\tau e^{-(2/3)s^{3/2}} \int^s e^{(2/3)t^{3/2}} \left[2t^{1/2} + K_1 - \frac{1}{2t} + O\left(\frac{1}{t^{5/2}}\right) \right] dt\,ds$$

$$\sim \int^\tau \left[2 + \frac{K_1}{s^{1/2}} + O\left(\frac{1}{s^{3/2}}\right) \right] ds \sim 2\tau + 2K_1\tau^{1/2} + K_2 + O\left(\frac{1}{\tau^{1/2}}\right).$$

We likewise obtain

$$w_3(\tau) \sim \frac{4}{3}\tau^{3/2} + 2K_1\tau + 2K_2\tau^{1/2} + K_3 + O\left(\frac{1}{\tau^{1/2}}\right).$$

Recall that, to order ϵ, the outer expansion is given by

$$Y_0(x) + \epsilon Y_1(x) \sim e^{2\sqrt{x}} \left[1 + \epsilon \left(-\frac{1}{2x} + \frac{2}{\sqrt{x}} - \frac{3}{2} \right) \right].$$

Rewriting this sum as a function of $\tau = x/\epsilon^{2/3}$, we obtain

$$Y_0 + \epsilon Y_1 \sim e^{2\epsilon^{1/3}\tau^{1/2}} \left[1 + \epsilon \left(-\frac{1}{2\tau\epsilon^{2/3}} + \frac{2}{\tau^{1/2}\epsilon^{1/3}} - \frac{3}{2} \right) \right]$$

$$= 1 + \epsilon^{1/3} \left(2\tau^{1/2} - \frac{1}{2\tau} \right) + \epsilon^{2/3} \left(2\tau + \frac{1}{\tau^{1/2}} \right)$$

$$+ \epsilon \left(\frac{4}{3}\tau^{3/2} + \frac{3}{2} \right) + O(\epsilon^{4/3}).$$

It will match the inner approximation

$$w_0(\tau) + \epsilon^{1/3} w_1(\tau) + \epsilon^{2/3} w_2(\tau) + \epsilon w_3(\tau)$$

through terms of order ϵ, for large τ, if we select

$$K_1 = 0, \qquad K_2 = 0, \qquad \text{and} \quad K_3 = \tfrac{3}{2}.$$

But, $K_1 = 0$ requires that $\lim_{\tau \to \infty} [w_1(\tau) - 2\tau^{1/2}] = 0$, so

$$c_1 = c_0 \lim_{\tau \to \infty} \left[2\tau^{1/2} - \int_0^\tau \int_0^s e^{-(2/3)(s^{3/2}-t^{3/2})} w_0(t)\,dt\,ds \right]$$

and w_1 also becomes completely specified. Likewise, we will obtain

$$c_2 = c_0 \lim_{\tau \to \infty} \left[2\tau - \int_0^\tau \int_0^s e^{-(2/3)(s^{3/2}-t^{3/2})} w_1(t)\,dt\,ds \right],$$

so w_2 is determined and

$$c_3 = c_0 \lim_{\tau \to \infty} \left[\frac{4}{3}\tau^{3/2} + \frac{3}{2} - \int_0^\tau \int_0^s e^{-(2/3)(s^{3/2}-t^{3/2})} w_2(t)\,dt\,ds \right].$$

Continuing in this fashion, we should be able to completely obtain all terms of the inner expansion successively. Likewise, the corresponding initial layer correction and composite expansion should be determined analogously.

F. Quasilinear Second-Order Problems

Consider the scalar problem

$$\epsilon\ddot{x} + f(x,t)\dot{x} + g(x,t) = 0, \qquad 0 \le t \le 1,$$

with $x(0)$ and $x(1)$ prescribed, and with smooth coefficients f and g. Based on our experience with linear problems, it seems natural to try to seek a solution with an $O(\epsilon)$ - thick initial layer of nonuniform convergence via the ansatz

$$x(t,\epsilon) = X(t,\epsilon) + u(\tau,\epsilon),$$

where the outer solution X has a power series expansion

$$X(t,\epsilon) \sim \sum_{j=0}^{\infty} X_j(t)\epsilon^j$$

and the initial layer correction $u(\tau,\epsilon)$ also has an expansion

$$u(\tau,\epsilon) \sim \sum_{j=0}^{\infty} u_j(\tau)\epsilon^j$$

whose terms decay to zero as the stretched variable $\tau = t/\epsilon$ tends to infinity. The outer expansion will then necessarily be a smooth solution of the terminal value problem

$$\epsilon\ddot{X} + f(X,t)\dot{X} + g(X,t) = 0, \qquad X(1,\epsilon) = x(1).$$

Equating terms successively implies that the limiting solution X_0 will satisfy the reduced problem

$$f(X_0,t)\dot{X}_0 + g(X_0,t) = 0, \qquad X_0(1) = x(1)$$

and succeeding X_j's will satisfy linear variational equations of the form

$$f(X_0,t)\dot{X}_j + f_x(X_0,t)X_j\dot{X}_0 + g_x(X_0,t)X_j = h_{j-1}(t), \qquad X_j(1) = 0,$$

where h_{j-1} is specified by preceding terms. In particular, if the reduced problem has a solution $X_0(t)$ throughout $0 \le t \le 1$ (which could be approximated numerically) along which

$$f(X_0(t),t) \neq 0,$$

successive terms will, in turn, be given by

$$X_j(t) = -\int_t^1 \exp\left[\int_t^r \frac{g_x(X_0(s),s)}{f(X_0(s),s)}\right.$$

$$\left. -\frac{f_x(X_0(s),s)}{f^2(X_0(s),s)}g(X_0(s),s))ds\right]\frac{h_{j-1}(r)}{f(X_0(r),r)}dr.$$

Thus, an outer expansion is readily obtained based on successive terminal value problems. Since the form of the assumed expansion implies that

$$\frac{d^j x}{dt^j} = \frac{d^j X}{dt^j} + \frac{1}{\epsilon^j}\frac{d^j u}{d\tau^j}$$

for $j = 0, 1$, and 2, the differential equation for x and X shows that the boundary layer correction u must satisfy

$$\frac{d^2 u}{d\tau^2} + f(X + u, \epsilon\tau)\frac{du}{d\tau} = \epsilon([f(X, \epsilon\tau) - f(X + u, \epsilon\tau)]\dot{X}$$
$$+ [g(X, \epsilon\tau) - g(X + u, \epsilon\tau)])$$

as a function of τ on $\tau \geq 0$, the initial condition $u(0, \epsilon) = x(0) - X(0, \epsilon)$, and u must decay to zero as $\tau \to \infty$. In particular, then, its leading term u_0 must satisfy the nonlinear initial value problem

$$\frac{d^2 u_0}{d\tau^2} + f(X_0(0) + u_0, 0)\frac{du_0}{d\tau} = 0, \qquad u_0(0) = x(0) - X_0(0),$$

whereas later terms will satisfy the linear variational problems

$$\frac{d^2 u_j}{d\tau^2} + f(X_0(0) + u_0, 0)\frac{du_j}{d\tau} + f_x(X_0(0) + u_0, 0)u_j\frac{du_0}{d\tau} = k_{j-1}(\tau),$$

$$u_j(0) = -X_j(0),$$

where k_{j-1} is a successively known function which decays exponentially toward zero as $\tau \to \infty$. Since the u_j's must also decay to zero, integration from infinity implies that the terms will in turn satisfy the first-order initial value problems

$$\frac{du_0}{d\tau} = -\int_0^{u_0} f(X_0(0) + s, 0)ds, \quad u_0(0) = x(0) - X_0(0),$$

and, for each $j > 0$,

$$\frac{du_j}{d\tau} + f(X_0(0) + u_0(\tau), 0)u_j = -\int_\tau^\infty k_{j-1}(s)ds, \quad u_j(0) = -X_j(0)$$

on $0 \leq \tau < \infty$. To guarantee the existence of decaying solutions u_j throughout $0 \leq t \leq 1$, we must add an appropriate *boundary layer stability assumption*. We will ask the sufficient condition that *there be a $\kappa > 0$ such that*

$$\frac{1}{\zeta}\int_0^\zeta f(X_0(0) + s, 0)ds \geq \kappa > 0$$

for all ζ between 0 and $u_0(0) = x(0) - X_0(0)$. Note that this condition requires that $f(x,0)$ have a positive average in the initial boundary layer. It also generally restricts the size of the boundary layer jump $|u_0(0)|$ and, in the special case when $f(x,t) = a(t)$, it simply requires that $a(0) > 0$. This and the earlier assumption that $f(X_0(t),t) \neq 0$ now require that

$$f(X_0(t),t) > 0 \qquad \text{throughout } 0 \leq t \leq 1.$$

(This results from applying l'Hôpital's rule in the limiting case that $\zeta \to 0$.) If $u_0(0) = 0, u_0(t) \equiv 0$. Otherwise, this assumption implies that $(1/u_0)(du_0/d\tau) \leq -\kappa$, so

$$0 \leq \frac{u_0(\tau)}{u_0(0)} \leq e^{-\kappa\tau} \quad \text{for all } \tau \geq 0$$

and $|u_0(\tau)|$ will decay exponentially to zero. In general, one has to obtain $u_0(\tau)$ numerically. Further, since $du_0/d\tau$ satisfies the homogeneous version of the equation for u_j, variation of parameters implies the unique solution

$$u_j(\tau) = \left(\int_0^{u_0} f(X_0(0) + s, 0)ds \right) \left[-\left(\int_0^{u_0(0)} f(X_0(0) + s, 0)ds \right)^{-1} X_j(0) \right.$$

$$\left. - \int_0^{\tau} \left(\int_0^{u_0(\sigma)} f(X_0(0) + t, 0)dt \right)^{-1} \int_{\sigma}^{\infty} k_{j-1}(s)ds d\sigma \right]$$

for each $j > 0$ which implies that the u_j's decay exponentially as $\tau \to \infty$. (Check it directly, if you wish!) Proofs that the formally constructed expansion is asymptotically valid are, more or less, given in Coddington and Levinson (1952), Chang and Howes (1984), and Smith (1985). We note that analogous methods could be used when x is a vector and the matrix $f(x,t)$ is positive definite [cf., e.g., Howes and O'Malley (1980)], though the details are more cumbersome.

Exercises

1. Show that the boundary value problem

$$\begin{cases} \epsilon\ddot{x} + e^x\dot{x} - \left(\dfrac{\pi}{2} \sin \dfrac{\pi t}{2} \right) e^{2x} = 0, \\ x(0) = x(1) = 0 \end{cases}$$

has a solution satisfying

$$x(t,\epsilon) = -\ln \left[\left(1 + \cos \frac{\pi t}{2} \right) \left(1 - \frac{1}{2}e^{-t/2\epsilon} \right) \right] + O(\epsilon).$$

2. Find two asymptotic solutions of

$$\begin{cases} \epsilon\ddot{x} + \dot{x} = (x-2)^2, \\[2mm] \dot{x}(0) - x(0) = 1, \qquad \dot{x}(1) + \tfrac{1}{3}x(1) = 1. \end{cases}$$

G. Existence, Uniqueness, and Numerical Computation of Solutions

Lorenz (1982a, b) considered the scalar quasilinear problem

$$\begin{cases} \epsilon u'' + f(u,x)u' + g(u,x) = 0, \qquad 0 \le x \le 1, \\[2mm] \text{with} \\[1mm] u(0) = A \quad\text{and}\quad u(1) = B \text{ prescribed} \end{cases}$$

for smooth coefficients f and g. When

$$g_u(u,x) \le f_x(u,x)$$

holds for all u and x values, he used inverse-monotonicity arguments to show that the problem has at most one solution. This relates to more familiar results obtainable via a maximum principle [cf. Protter and Weinberger (1967)]. Dorr et al. (1973), for example, showed that for

$$g(u,x) = h(u,x)u + j(u,x),$$

where

$$h(u,x) \le -\delta < 0$$

everywhere, the solution u satisfies

$$\max_{0 \le x \le 1} |u(x)| \le \max(|A|,|B|) + \frac{1}{\delta}\max|j(u,x)|.$$

This result can often be used to prove uniqueness by showing that the difference between any two possible solutions is trivial. Lorenz also used compactness arguments to prove the existence of a unique solution $u(x)$, when $f_x \equiv 0$, assuming either that

$$g_u(u,x) \le -\delta < 0$$

or

$$g_u(u,x) \le 0 \quad\text{and}\quad g(0,x) = 0$$

holds everywhere. Such existence results can also often be obtained by explicitly finding smooth upper and lower solutions $\beta(x)$ and $\alpha(x)$ which satisfy the differential inequalities

$$\epsilon\beta'' + f(\beta, x)\beta' + g(\beta, x) \leq 0$$

and

$$\epsilon\alpha'' + f(\alpha, x)\alpha' + g(\alpha, x) \geq 0$$

throughout $0 \leq x \leq 1$ as well as the endpoint inequalities $\alpha(0) \leq A \leq \beta(0)$ and $\alpha(1) \leq B \leq \beta(1)$. Then, a more general result of Nagumo (1937), as presented in Jackson (1968) and Bernfeld and Lakshmikantham (1974), shows that the boundary value problem has a smooth solution u satisfying

$$\alpha(t) \leq u(t) \leq \beta(t)$$

throughout [0,1]. Howes (1976, 1978), Chang and Howes (1984), and Weinstein and Smith (1975) give numerous examples (including some where the differential equation is quadratic in u') showing how clever adaptations of our familiar asymptotic constructions can result in asymptotically tight bounding functions α and β.

As a result of much computational experience, those active in scientific computation have learned that one way to solve these difficult two-point problems numerically (at least when f is independent of x) is to obtain their solutions as steady states of the admittedly artificially posed initial boundary value problem for the parabolic equation

$$u_t = \epsilon u_{xx} + f(u)u_x + g(u, x).$$

When $g_u \leq 0$ and $g(0, x) = 0$, for example, Osher (1981) showed convergence of the numerical solution to a unique steady state by using appropriate one-sided difference approximations for u_x determined by the sign of f. This strong result is fascinating, because the solution is not prevented from passing through any number and kind of turning points, i.e. zeros of f. In many applied fields, but especially in meteorology, it has long been common practice to use such "upwinding" techniques. Lorenz (1982a,b, 1984) studied the two-point problem for

$$\epsilon u'' + f(u)u' + g(u, x) = 0$$

directly, assuming that $g_u \leq -\delta < 0$, and proving that a unique limiting solution of bounded variation exists, which could only be discontinuous at the endpoints or at turning points. The difference scheme he used [for an $O(h)$ error estimate] is given by

$$\frac{\epsilon}{h^2}(u_{i+1} - 2u_i + u_{i-1}) + \frac{1}{h}[k(u_{i+1}, u_i) - k(u_i, u_{i-1})] + g(u_i, ih) = 0,$$

where $u_i = u(ih)$, $i = 1, 2, \ldots, m$, $h = 1/(m+1)$ and (among other possible choices) he selected the Engquist–Osher method which uses

$$k(u, v) = \int\limits_0^u \min[0, f(s)]ds + \int\limits_0^v \max[0, f(s)]ds,$$

even for $\epsilon = 0$. Further characterizations of the limiting solution [cf. Lorenz (1984)] are analogous to results already obtained by constructing appropriate endpoint and interior layers, e.g., that the limiting solution u satisfies the reduced equation away from points of discontinuity, that the auxiliary conditions which specify u on such intervals are determined by the sign of $f(u)$, and that inequalities concerning the sign of $F(y) = \int_{u(x)}^{y} f(s)ds$ hold near jump points and in endpoint layers.

Numerical solutions can, of course, be obtained by numerical evaluation of asymptotic solutions [cf. e.g., Flaherty and O'Malley (1977, 1984)]. We would, however, especially prefer to obtain numerical approximations to solutions in situations where asymptotic solutions are not available. Instead of using a constant mesh, like Osher, it is natural to move mesh points to regions where solutions of such singular perturbation problems change rapidly. [Flaherty et al. (1989) discuss such methods for partial differential equations.] This is, of course, quite analogous to the analytic use of stretching transformations (or of equidistributing appropriate functions, such as arc length, to determine a mesh). Kreiss et al. (1986), among others, have studied linear systems of such differential equations using a combination of symmetric and unsymmetric difference equations, while Ascher et al. (1988) emphasize the use of a "theoretical multiple shooting" technique which is based on dichotomy concepts and decoupling of rapidly increasing and rapidly decreasing solution modes. For nonlinear problems, one would use Newton's method, linearizing about a reasonable guess of a solution [cf. Cash and Wright (1989)]. The book [Ascher et al. (1988)] nicely illustrates the current vitality of this active research area, and it also demonstrates the many challenges which lie ahead for the numerical analyst who tackles applied singular perturbation problems. A large collection of papers is surveyed in Kadalbajoo and Reddy (1989), though not in depth.

H. Quasilinear Vector Problems

The approach used to solve linear boundary value problems with endpoint layers also applies to some nonlinear problems and, in particular, to certain systems which are linear in the fast variable. Let us, then, consider the quasilinear system

$$\begin{cases} \dot{x} = A(x,t) + B(x,t)y, \\ \epsilon\dot{y} = C(x,t) + D(x,t)y \end{cases}$$

of $m + n$ differential equations on the interval $0 \leq t \leq 1$ subject to r initial and s terminal conditions

$$L(x(0), y(0)) = 0$$

and

$$Q(x(1), y(1)) = 0,$$

where $r + s = m + n$. Let us further assume that the $n \times n$ matrix $D(x, t)$ has a hyperbolic splitting, with k stable and $n - k$ unstable eigenvalues for all x and all t in [0,1]. (This hypothesis, regrettably, eliminates all turning points.) Let us also assume that at least k initial conditions and at least $n - k$ terminal conditions are prescribed, i.e., that

$$r \geq k \quad \text{and} \quad s \geq n - k.$$

[A special case arises from the first (or Dirichlet) boundary value problem for the second-order vector equation $\epsilon \ddot{z} + f(z, t)\dot{z} + g(z, t) = 0$ where f is a positive or negative definite matrix.] We might naturally seek an asymptotic solution to our problem in the form

$$\begin{cases} x(t, \epsilon) = X(t, \epsilon) + \epsilon \xi(\tau, \epsilon) + \epsilon \eta(\sigma, \epsilon), \\ y(t, \epsilon) = Y(t, \epsilon) + \zeta(\tau, \epsilon) + \theta(\sigma, \epsilon) \end{cases}$$

with an outer solution $\binom{X}{Y}$ having a power series expansion

$$\begin{pmatrix} X(t, \epsilon) \\ Y(t, \epsilon) \end{pmatrix} \sim \sum_{j=0}^{\infty} \begin{pmatrix} X_j(t) \\ Y_j(t) \end{pmatrix} \epsilon^j$$

in ϵ, an initial layer correction $\binom{\epsilon \xi}{\zeta}$ having an expansion

$$\begin{pmatrix} \xi(\tau, \epsilon) \\ \zeta(\tau, \epsilon) \end{pmatrix} \sim \sum_{j=0}^{\infty} \begin{pmatrix} \xi_j(\tau) \\ \zeta_j(\tau) \end{pmatrix} \epsilon^j$$

whose terms tend to zero as $\tau = t/\epsilon \to \infty$, and a terminal layer correction $\binom{\epsilon \eta}{\theta}$ with an expansion

$$\begin{pmatrix} \eta(\sigma, \epsilon) \\ \theta(\sigma, \epsilon) \end{pmatrix} \sim \sum_{j=0}^{\infty} \begin{pmatrix} \eta_j(\sigma) \\ \theta_j(\sigma) \end{pmatrix} \epsilon^j$$

which decays to zero as $\sigma = (1 - t)/\epsilon \to \infty$. [For $k = 0$ (or n), we naturally omit the initial (or terminal) layer corrections.] Thus, we are seeking a solution where x converges uniformly throughout $0 \leq t \leq 1$ and where y converges within the interval and remains bounded, but has endpoint regions of rapid change. {If we allowed y to have endpoint impulses, solutions with other asymptotic behaviors might be possible [cf., e.g., O'Malley (1970b) and (1980)].}

Within (0,1), the outer solution is then a smooth solution of the system, so

$$\begin{cases} \dot{X} = A(X, t) + B(X, t)Y, \\ \epsilon \dot{Y} = C(X, t) + D(X, t)Y \end{cases}$$

must be satisfied as a power series in ϵ. Thus, the leading term $\begin{pmatrix} X_0 \\ Y_0 \end{pmatrix}$ must satisfy the differential-algebraic system

$$\begin{cases} \dot{X}_0 = A(X_0, t) + B(X_0, t)Y_0, \\ 0 = C(X_0, t) + D(X_0, t)Y_0, \end{cases}$$

whereas later terms will satisfy a linearized system

$$\begin{cases} \dot{X}_j = A_x(X_0, t)X_j + B_x(X_0, t)X_jY_0 + B(X_0, t)Y_j + \gamma_{j-1}(t), \\ 0 = C_x(X_0, t)X_j + D_x(X_0, t)X_jY_0 + D(X_0, t)Y_j + \delta_{j-1}(t), \end{cases}$$

where the nonhomogeneous terms γ_{j-1} and δ_{j-1} are known successively in terms of preceding coefficients. Since $D(X_0, t)$ is nonsingular, we will have

$$Y_0 = -D^{-1}(X_0, t)C(X_0, t)$$

and

$$Y_j = -D^{-1}(X_0, t)[C_x(X_0, t)X_j - D_x(X_0, t)X_jD^{-1}(X_0, t)C(X_0, t) + \delta_{j-1}(t)]$$

for $j > 0$, leaving us the limiting nonlinear system

$$\dot{X}_0 = A(X_0, t) - B(X_0, t)D^{-1}(X_0, t)C(X_0, t)$$

and the successive nonhomogeneous linear (variational) equations

$$\dot{X}_j = [A_x(X_0, t) - B(X_0, t)D^{-1}(X_0, t)C_x(X_0, t)]X_j$$
$$- [B_x(X_0, t) - B(X_0, t)D^{-1}(X_0, t)D_x(X_0, t)]X_jD^{-1}(X_0, t)C(X_0, t)$$
$$+ [\gamma_{j-1}(t) - B(X_0, t)D^{-1}(X_0, t)\delta_{j-1}(t)].$$

To successively determine the terms X_j of the outer expansion, then, we need only specify m appropriate boundary conditions and be assured that the resulting two-point problem for each X_j on $0 \le t \le 1$ is solvable. Since $X(0, \epsilon)$ is generally unspecified, it is useful to realize that we have obtained a smooth m-dimensional manifold of outer solutions parameterized by this initial vector.

We note that the assumed form of solution implies that

$$\frac{dx}{dt} = \frac{dX}{dt} + \frac{d\xi}{d\tau} - \frac{d\eta}{d\sigma}$$

and that

$$\epsilon\frac{dy}{dt} = \epsilon\frac{dY}{dt} + \frac{d\zeta}{d\tau} - \frac{d\theta}{d\sigma}.$$

Near $x = 0$ (or, respectively, $x = 1$), the functions of σ (or τ) are asymptotically negligible, so the initial layer correction $\begin{pmatrix} \epsilon\xi \\ \zeta \end{pmatrix}$ must provide a decaying solution of the initial layer system

$$\frac{d\xi}{d\tau} = B(X + \epsilon\xi, \epsilon\tau)\zeta + [B(X + \epsilon\xi, \epsilon\tau) - B(X, \epsilon\tau)]Y$$

$$+ [A(X + \epsilon\xi, \epsilon\tau) - A(X, \epsilon\tau)],$$

$$\frac{d\zeta}{d\tau} = D(X + \epsilon\xi, \epsilon\tau)\zeta + [D(X + \epsilon\xi, \epsilon\tau) - D(X, \epsilon\tau)]Y$$

$$+ [C(X + \epsilon\xi, \epsilon\tau) - C(X, \epsilon\tau)].$$

The terminal layer correction will likewise be determined through the system

$$\frac{d\eta}{d\sigma} = -B(X + \epsilon\eta, 1 - \epsilon\sigma)\theta + [B(X, 1 - \epsilon\sigma) - B(X + \epsilon\eta, 1 - \epsilon\sigma)]Y$$

$$+ [A(X, 1 - \epsilon\sigma) - A(X + \epsilon\eta, 1 - \epsilon\sigma)],$$

$$\frac{d\theta}{d\sigma} = -D(X + \epsilon\eta, 1 - \epsilon\sigma)\theta + [D(X, 1 - \epsilon\sigma) - D(X + \epsilon\eta, 1 - \epsilon\sigma)]Y$$

$$+ [C(X, 1 - \epsilon\sigma) - C(X + \epsilon\eta, 1 - \epsilon\sigma)].$$

Because the original system is linear in the fast variable y, the limiting initial layer system

$$\begin{cases} \dfrac{d\xi_0}{d\tau} = B(X_0(0), 0)\zeta_0, \\[2mm] \dfrac{d\zeta_0}{d\tau} = D(X_0(0), 0)\zeta_0 \end{cases}$$

is linear, and even has a constant system matrix. The decay requirement on ζ_0 forces us to take

$$\zeta_0(\tau) = e^{D(X_0(0),0)\tau} P_0(X_0(0))\zeta_0(0)$$

where we have introduced a matrix P_0 which projects onto the k-dimensional stable eigenspace of $D(X_0(0), 0)$. Likewise, the decay of ξ_0 requires us to take

$$\xi_0(\tau) = -B(X_0(0), 0) \left(\int_\tau^\infty e^{D(X_0(0),0)s} ds \right) P_0(X_0(0))\zeta_0(0)$$

$$= B(X_0(0), 0)D^{-1}(X_0(0), 0)\zeta_0(\tau).$$

Later terms of this initial layer correction satisfy nonhomogeneous systems

$$\begin{cases} \dfrac{d\xi_j}{d\tau} = B(X_0(0), 0)\zeta_j + \alpha_{j-1}(\tau), \\[2mm] \dfrac{d\zeta_j}{d\tau} = D(X_0(0), 0)\zeta_j + \beta_{j-1}(\tau), \end{cases}$$

where, by induction, the α_{j-1} and β_{j-1} terms are successively determined, exponentially decaying vectors which lie in the stable eigenspace of $D(X_0(0), 0)$. Integrating, then, the decaying solutions $\binom{\xi_j}{\zeta_j}$ must have the form

$$
\begin{cases}
\zeta_j(\tau) = e^{D(X_0(0),0)\tau} P_0(X_0(0))\zeta_j(0) + \int\limits_0^\tau e^{D(X_0(0),0)(\tau-s)} \beta_{j-1}(s)ds \\[2em]
\xi_j(\tau) = -\int\limits_\tau^\infty \frac{d\xi_j}{d\tau}(r)dr.
\end{cases}
$$

Both these vectors lie in the range of P_0. Thus, the initial layer correction $\binom{\epsilon\xi}{\zeta}$ will be completely specified asymptotically once we provide the initial value $P_0(X_0(0))\zeta(0, \epsilon)$. Hence, we have determined a k manifold of rapidly decaying initial layer corrections, parameterized by the limiting outer solution value $X_0(0)$. Analogously, the terminal layer correction will become determined upon specification of

$$P_1(X_0(1))\theta(0, \epsilon),$$

where P_1 is the corresponding projection onto the $(n-k)$-dimensional unstable eigenspace of $D(X_0(1), 1)$, i.e., the $(n-k)$-dimensional manifold of fast-growing layer solutions is parameterized by $X_0(1)$.

Since $x(0) \sim X(0, \epsilon) + \epsilon\xi(0, \epsilon)$ and $y(0) \sim Y(0, \epsilon) + P_0(X_0(0))\zeta(0, \epsilon)$, the r initial conditions take the asymptotic form

$$L(X(0, \epsilon) + \epsilon\xi(0, \epsilon), Y(0, \epsilon) + P_0(X_0(0))\zeta(0, \epsilon)) = 0.$$

The terminal conditions analogously yield

$$Q(X(1, \epsilon) + \epsilon\eta(0, \epsilon), Y(1, \epsilon) + P_1(X_0(1))\theta(0, \epsilon)) = 0.$$

When $\epsilon = 0$, then, we obtain the r limiting initial conditions

$$L(X_0(0), -D^{-1}(X_0(0), 0)C(X_0(0), 0) + P_0(X_0(0))\zeta_0(0)) = 0$$

and the s analogous terminal conditions

$$Q(X_0(1), -D^{-1}(X_0(1), 1)C(X_0(1), 1) + P_1(X_0(1))\theta_0(0)) = 0.$$

Now *assume* that k of these r initial conditions can be solved to obtain

$$P_0(X_0(0))\zeta_0(0) \equiv \psi_0(X_0(0))$$

and that $n-k$ of these s terminal conditions can be solved for

$$P_1(X_0(1))\theta_0(0) \equiv \phi_0(X_0(1)).$$

Note that we do allow multiple solutions, if they are isolated. The reduced problem will now be defined by the reduced equation, the remaining $r - k$ limiting initial conditions, and the remaining $s - n + k$ limiting terminal conditions. (We have obtained a cancellation law for our quasilinear problem!) Thus, we naturally *assume* that the reduced problem

$$\begin{cases} \dot{X}_0 = A(X_0, t) - B(X_0, t)D^{-1}(X_0, t)C(X_0, t), \quad 0 \le t \le 1, \\ L(X_0(0), -D^{-1}(X_0(0), 0)C(X_0(0), 0) + \psi_0(X_0(0))) = 0, \\ Q(X_0(1), -D^{-1}(X_0(1), 1)C(X_0(1), 1) + \phi_0(X_0(1))) = 0 \end{cases}$$

has an isolated (but not necessarily unique) solution $X_0(t)$ on $0 \le t \le 1$. (Here, of course, k initial conditions and $n - k$ terminal conditions will be trivially satisfied, due to the definitions of ψ_0 and ϕ_0.) Note that selecting any such $X_0(t)$ uniquely specifies a corresponding $Y_0(t)$.

Higher-order terms in the formal expansion can then be obtained successively. Thus, the ϵ^j coefficients in the boundary conditions imply that we must satisfy linear equations

$$\mathcal{L}(X_j(0), P_0(X_0(0))\zeta_j(0)) \equiv L_x(X_0(0), Y_0(0) + \psi_0(X_0(0)))X_j(0)$$
$$+ L_y(X_0(0), Y_0(0) + \psi_0(X_0(0)))(Y_j(0)$$
$$+ P_0(X_0(0))\zeta_j(0)) = \kappa_{j-1}$$

and

$$\mathcal{Q}(X_j(1), P_1(X_1(1))\theta_j(0)) \equiv Q_x(X_0(1), Y_0(1) + \phi_0(X_0(1)))X_j(1)$$
$$+ Q_y(X_0(1), Y_0(1) + \phi_0(X_0(1)))(Y_j(1)$$
$$+ P_1(X_0(1))\theta_j(0)) = \lambda_{j-1}$$

where κ_{j-1} and λ_{j-1} are successively known. If we now *assume* that

$$\begin{cases} L_y(X_0(0), Y_0(0) + \psi_0(X_0(0)))P_0(X_0(0)) \\ \quad \text{has constant rank } k \text{ (for all } X_0(0)) \quad \text{and} \\ Q_y(X_0(1), Y_0(1) + \phi_0(X_0(1)))P_1(X_0(1)) \\ \quad \text{likewise has constant rank } n - k, \end{cases}$$

we will be able to uniquely solve k of the initial conditions to obtain

$$P_0(X_0(0))\zeta_j(0) \equiv \psi(X_0(0))X_j(0) + \ell_{j-1}$$

and, analogously, to solve $n - k$ of the terminal conditions to get

$$P_1(X_0(1))\theta_j(0) \equiv \phi(X_0(1))X_j(1) + m_{j-1}.$$

Here, ℓ_{j-1} and m_{j-1} are known successively. It follows that the jth coefficient X_j in the outer expansion will be specified as the solution of an mth-order linear system, subject to $r-k$ initial conditions and $s-n+k$ terminal conditions. It will have a unique solution *provided* the corresponding linear homogeneous boundary value problem

$$\begin{cases} \dot{z} = [A_x(X_0,t) - B(X_0,t)D^{-1}(X_0,t)C_x(X_0,t)]z - [B_x(X_0,t) \\ \qquad - B(X_0,t)D^{-1}(X_0,t)D_x(X_0,t)]zD^{-1}(X_0,t)C(X_0,t), \\ \psi(X_0(0))z(0) = 0, \\ \phi(X_0(1))z(1) = 0 \end{cases}$$

has only the trivial solution $z(x) \equiv 0$.

Under the accumulated (but natural!) assumptions, we have been able to construct a formal solution to our boundary value problem corresponding to every solution $X_0(t)$ of the reduced problem. A relatively straightforward proof of asymptotic correctness can be obtained by many linearization and Green's function arguments [cf. Vasil'eva and Butuzov (1973), O'Malley (1980), Schmeiser and Weiss (1986), and Jeffries and Smith (1989)].

Example 1

Consider the quasilinear system

$$\begin{cases} \dot{x} = -x + xy, \\ \epsilon\dot{y} = x - (x+1)y \end{cases}$$

subject to the (unseparated) two-point boundary conditions

$$x(0) = 1, \qquad y^2(0) = y(1).$$

The solution of the corresponding reduced problem

$$\dot{X}_0 = -X_0 + X_0Y_0, \qquad X_0(0) = 1,$$

$$0 = X_0 - (X_0 + 1)Y_0$$

is given by

$$Y_0 = \frac{X_0}{1 + X_0}$$

provided the unique solution $X_0(t)$ of the resulting initial value problem

$$\dot{X}_0 = -\frac{X_0}{1 + X_0}, \qquad X_0(0) = 1$$

does not become -1. We note that the solution $X_0 + \ln X_0 = 1 - t$ could be easily computed numerically and that it decreases monotonically to zero. Indeed, since $\frac{1}{2} \le 1/(1 + X_0) \le 1$, we will have

$$e^{-t} \le X_0(t) \le e^{-t/2}.$$

The negative sign of the Jacobian $-(X_0 + 1)$ suggests that there will be an initial layer in the fast variable y, so we seek a solution of the form

$$x(t) = X_0(t) + O(\epsilon),$$

$$y(t) = Y_0(t) + \zeta_0(\tau) + O(\epsilon),$$

where $\zeta_0 \to 0$ as $\tau = t/\epsilon \to \infty$. Thus, ζ_0 will necessarily satisfy

$$\frac{d\zeta_0}{d\tau} = -(X_0(0) + 1)\zeta_0,$$

so

$$\zeta_0(\tau) = e^{-2\tau}\zeta_0(0).$$

The boundary condition for y implies that $[Y_0(0) + \zeta_0(0)]^2 = Y_0(1)$, so

$$\zeta_0(0) = -\frac{1}{2} \pm \sqrt{\frac{X_0(1)}{1 + X_0(1)}}.$$

Thus, there are two possible initial layer corrections $\zeta_0(\tau)$ corresponding to the same limiting solution $X_0(t)$. It is straightforward to determine higher-order asymptotic approximations corresponding to each of them and to indeed show that the boundary value problem has two solutions for any sufficiently small ϵ.

Example 2

Consider the quasilinear system

$$\begin{cases} \dot{x} = 1 - x, \\[2mm] \epsilon \dot{y}_1 = y_2, \\[2mm] \epsilon \dot{y}_2 = \alpha^2(x)y_1 + \beta(x), \end{cases}$$

where $\alpha(x) = 1 + 2x$ and $\beta(x) = 8x(1 - x)$. As long as $\alpha(x)$ is nonzero, the coefficient matrix for the fast system will have positive and negative eigenvalues $\pm|\alpha(x)|$, so we are led to seek an asymptotic solution of the form

$$x(t) = X_0(t) + O(\epsilon),$$

$$y_i(t) = Y_{i0}(t) + \zeta_{i0}(\tau) + \theta_{i0}(\sigma) + O(\epsilon), \quad i = 1, 2,$$

where the $\zeta_{i0} \to 0$ as $\tau = t/\epsilon \to \infty$ and $\theta_{i0} \to 0$ as $\sigma = (1 - t)/\epsilon \to \infty$. Proceeding as usual, we obtain

$$X_0(t) = 1 + [x(0) - 1]e^{-t},$$

$$Y_{10}(t) = -\beta(X_0(t))/\alpha^2(X_0(t)) \text{ provided } \alpha(X_0) \text{ remains nonzero,}$$

and

$$Y_{20}(t) = 0.$$

The limiting initial layer correction will satisfy

$$\begin{cases} \dfrac{d\zeta_{10}}{d\tau} = \zeta_{20}, \\[2mm] \dfrac{d\zeta_{20}}{d\tau} = \alpha^2(x(0))\zeta_{10}, \end{cases}$$

so $d^2\zeta_{10}/d\tau^2 = \alpha^2(x(0))\zeta_{10}$ implies that the decaying solution is given by

$$\zeta_{10}(\tau) = e^{-|\alpha(x(0))|\tau}\zeta_{10}(0)$$

and

$$\zeta_{20}(\tau) = -|\alpha(x(0))|\zeta_{10}(\tau).$$

Analogously, we obtain

$$\theta_{10}(\sigma) = e^{-|\alpha(X_0(1))|\sigma}\theta_{10}(0)$$

and

$$\theta_{20}(\sigma) = |\alpha(X_0(1))|\theta_{10}(\sigma),$$

leaving only the constants $x(0), \zeta_{10}(0),$ and $\theta_{10}(0)$ to be determined. Now suppose we were given the separated boundary conditions

$$\begin{cases} x(0) + y_1(0) = 0, \\[2mm] 2x(0) - y_2(0) = 0, \\[2mm] x(1) + y_1(1) = 0. \end{cases}$$

The first and third conditions imply that

$$\zeta_{10}(0) = \frac{\beta(x(0))}{\alpha^2(x(0))} - x(0),$$

and

$$\theta_{10}(0) = \frac{\beta(X_0(1))}{\alpha^2(X_0(1))} - X_0(1),$$

whereas the second condition provides us the polynomial

$$2x(0) = -|\alpha(x(0))| \left(\frac{\beta(x(0))}{\alpha^2(x(0))} - x(0) \right)$$

to determine $x(0)$. It has the three roots

$$x(0) = 0, \qquad -1 + \tfrac{1}{2}\sqrt{13}, \quad \text{and} \ -2 - \tfrac{1}{2}\sqrt{21}$$

which determine three functions $X_0(t)$ along which $\alpha(X_0(t)) \neq 0$ on $0 \leq t \leq 1$. We are guaranteed, then, that the boundary value problem has at least three solutions for ϵ sufficiently small, corresponding to the three different limiting solutions $\binom{X_0(t)}{Y_0(t)}$. We anticipate that any other solutions would have a different asymptotic structure.

I. An Example with an Angular Solution

Haber and Levinson (1955) developed a theory for scalar two-point problems for equations

$$\epsilon y'' = f(x, y, y'),$$

where f is nonlinear in y' and $y(0)$ and $y(1)$ are prescribed [cf. also Howes (1978)]. Under appropriate conditions, solutions of the two reduced problems

$$\begin{cases} f(x, Y_{L0}, Y'_{L0}) = 0, & Y_{L0}(0) = y(0), \\ f(x, Y_{R0}, Y'_{R0}) = 0, & Y_{R0}(1) = y(1) \end{cases}$$

can intersect at an interior point, allowing the possibility of a continuous limiting solution within the interval.

A typical example is provided by

$$\begin{cases} \epsilon y'' = 1 - (y')^2, \\ y(0) = 0, & y(1) = \tfrac{1}{2}. \end{cases}$$

[The boundary conditions could be changed to any values $y(0)$ and $y(1)$ with $|y(0) - y(1)| < 1$, but for other boundary values, limiting solutions would necessarily be different.] The maximum principle guarantees a unique solution. Moreover, the reduced equation

$$1 - (Y'_0)^2 = 0$$

and the boundary conditions suggest the possible limiting solutions

$$\begin{cases} Y_{L0}^{\pm}(x) = \pm x, & 0 \leq x \leq \tilde{x}^{\pm}, \\ Y_{R0}^{\pm}(x) = \pm(1 - x) + \tfrac{1}{2}, & \tilde{x}^{\pm} \leq x \leq 1. \end{cases}$$

We note that $Y_{L0}^{+}(x)$ and $Y_{R0}^{+}(x)$ intersect at $\tilde{x}_1 = \tfrac{3}{4}$, while $Y_{L0}^{-}(x)$ and $Y_{R0}^{-}(x)$ intersect at $\tilde{x}_2 = \tfrac{1}{4}$. Pictorially, we have Figures 3.3 – 3.6. Note that the derivative $Y_0^{\pm}(x)$ jumps at $\tilde{x}$. This suggests the possibility of constructing asymptotic solutions of the form

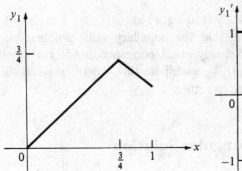

Fig. 3.3. The solution y_1.

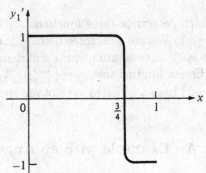

Fig. 3.4. The derivative y_1'.

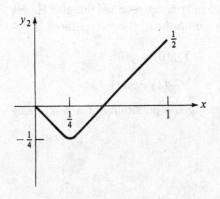

Fig. 3.5. The solution y_2.

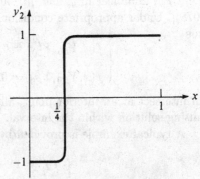

Fig. 3.6. The derivative y_2'.

$$y(x) = \begin{cases} Y_L(x,\epsilon) + \epsilon\xi(\kappa,\epsilon), & 0 \le x \le \tilde{x}, \\ Y_R(x,\epsilon) + \epsilon\eta(\kappa,\epsilon), & \tilde{x} \le x \le 1, \end{cases}$$

where

$$\kappa = (x - \tilde{x})/\epsilon$$

and $\xi \to 0$ as $\kappa \to -\infty$, whereas $\eta \to 0$ as $\kappa \to \infty$. Continuity of y and y' at $\tilde{x}$ will require that

$$Y_L(\tilde{x},\epsilon) + \epsilon\xi(0,\epsilon) = Y_R(\tilde{x},\epsilon) + \epsilon\eta(0,\epsilon)$$

while

$$Y_L'(\tilde{x},\epsilon) + \frac{d\xi}{d\kappa}(0,\epsilon) = Y_R'(\tilde{x},\epsilon) + \frac{d\eta}{d\kappa}(0,\epsilon).$$

Since the outer solutions Y_L and Y_R must satisfy the differential equation as a power series in ϵ, we easily show that

$$Y_L^\pm(x, \epsilon) = Y_{L0}^\pm(x) = \pm x$$

and

$$Y_R^\pm(x, \epsilon) = Y_{R0}^\pm(x) = \pm(1 - x) + \tfrac{1}{2}.$$

Then, $y' = Y_{L0}'^\pm(x) + d\xi^\pm/d\kappa = \pm 1 + d\xi^\pm/d\kappa$ and $\epsilon y'' = d^2\xi^\pm/d\kappa^2$ for $x \le \tilde{x}$ imply that $\xi^\pm$ must satisfy

$$\frac{d^2\xi^\pm}{d\kappa^2} = 1 - \left(\pm 1 + \frac{d\xi^\pm}{d\kappa}\right)^2 = \mp 2\frac{d\xi^\pm}{d\kappa} - \left(\frac{d\xi^\pm}{d\kappa}\right)^2$$

for $\kappa \le 0$. Since we expect $-1 \le y' \le 1$, we will have $-1 \mp 1 \le d\xi^\pm/d\kappa \le 1 \mp 1$. In particular, then, $a = d\xi^+/d\kappa$ should satisfy

$$\frac{da}{d\kappa} = -a(a+2) \text{ on } \kappa \le 0 \text{ with } -2 < a(0) < 0.$$

Any such solution would, however, have the limit $a(-\infty) = -2$ as $\kappa \to -\infty$, contradicting the requirement that $a(-\infty) = 0$. For $b = d\xi^-/d\kappa$, however, we would instead have

$$\frac{db}{d\kappa} = b(2 - b) \text{ on } \kappa \le 0 \text{ with } 0 < b(0) < 2.$$

This has the desired limit $b(-\infty) = 0$, suggesting that y^- provides the appropriate asymptotic solution as $\epsilon \to 0$. Indeed, a direct integration yields

$$b(\kappa) = \frac{d\xi^-}{d\kappa} = \frac{2b(0)e^{2\kappa}}{2 - b(0) + b(0)e^{2\kappa}},$$

so

$$\xi^-(\kappa) = \int_{-\infty}^{\kappa} \frac{d\xi^-}{d\kappa}(s)\,ds = \ln\left(1 + \frac{b(0)}{2 - b(0)}e^{2\kappa}\right)$$

on $\kappa \le 0$. Proceeding analogously for $\kappa > 0$, we obtain

$$\eta^+(\kappa) = \ln\left(1 + \frac{2 - b(0)}{b(0)}e^{-2\kappa}\right)$$

[having matched $y'(\tilde{x}^\pm)$]. In order for y^- to be continuous at $\tilde{x}$, we must have $\xi^-(0) = \eta^+(0)$, i.e., $b(0) = 1$. Thus, we have found a unique asymptotic solution such that

$$y(x, \epsilon) \sim \begin{cases} -x + \epsilon \ln(1 + e^{(2/\epsilon)(x-1/4)}), & 0 \le x \le \tfrac{1}{4}, \\ x - \tfrac{1}{2} + \epsilon \ln(1 + e^{-(2/\epsilon)(x-1/4)}), & \tfrac{1}{4} \le x \le 1. \end{cases}$$

Note that $y(\tfrac{1}{4}, \epsilon) \sim -\tfrac{1}{4} + \epsilon \ln 2$, that the expansion for $y'(x, \epsilon)$ is also determined, and that $y'(\tfrac{1}{4}, \epsilon) \sim 0$.

J. Nonlinear Systems

Consider the singularly perturbed nonlinear system

$$\begin{cases} \dot{x} = f(x, y, t, \epsilon), \\ \epsilon \dot{y} = g(x, y, t, \epsilon) \end{cases}$$

of $m + n$ differential equations subject to $m + n$ boundary conditions

$$a(x(0), y(0), x(1), y(1), \epsilon) = 0.$$

We will assume that f, g, and a are smooth functions of their arguments. As we have already found, solutions may have interior locations where y or its derivatives converge nonuniformly as $\epsilon \to 0$. Still, in many applications, only endpoint layers have primary importance. Thus, it is still natural to attempt to obtain asymptotic solutions such that

$$x(t, \epsilon) = X_0(t) + O(\epsilon),$$

$$y(t, \epsilon) = Y_0(t) + \mu_0(\tau) + v_0(\sigma) + O(\epsilon),$$

where $\mu_0 \to 0$ as $\tau = t/\epsilon \to \infty$ and $v_0 \to 0$ as $\sigma = (1 - t)/\epsilon \to \infty$. Then, the slowly varying x vector will have a uniform limit $X_0(t)$ throughout $0 \leq t \leq 1$, whereas the fast vector y will be bounded, it will change rapidly in an $O(\epsilon)$ thick boundary layer region near each endpoint, and it will be slowly varying within the interval. We can, indeed, expect such a solution structure to apply along any limiting solution $\binom{X_0(t)}{Y_0(t)}$ when the Jacobian matrix

$$g_y(X_0(t), Y_0(t), t, 0)$$

remains nonsingular and maintains a fixed number of eigenvalues in each half-plane and the boundary conditions are compatible. When this ansatz is appropriate, the limiting solution will necessarily satisfy the reduced system

$$\begin{cases} \dot{X}_0 = f(X_0, Y_0, t, 0), \\ 0 = g(X_0, Y_0, t, 0). \end{cases}$$

As for the corresponding initial value problem, one could then determine the root

$$Y_0(t) = \varphi(X_0(t), t)$$

of $g = 0$ by integrating the system

$$\dot{Y}_0(t) = -g_y^{-1}(X_0, Y_0, t, 0)[g_x(X_0, Y_0, t, 0)f(X_0, Y_0, t, 0) + g_t(X_0, Y_0, t, 0)]$$

(obtained by differentiating the algebraic equation $g = 0$) together with the reduced-order system

$$\dot{X}_0 = f(X_0, \varphi(X_0, t), t, 0).$$

The appropriate initial vector $X_0(0)$ generally remains unspecified. If we knew it, we could determine the terminal value $\binom{X_0(1)}{Y_0(1)}$ of the outer solution by integration (presuming continued existence of $X_0(t)$ on $0 \le t \le 1$). Likewise, the differential systems satisfied by the boundary layer corrections μ_0 and v_0 would then follow since we must have

$$\epsilon \frac{dy}{dt} \sim \epsilon \frac{dY_0}{dt} + \frac{d\mu_0}{d\tau} \approx \frac{d\mu_0}{d\tau}$$

for $t < 1$ and $\epsilon\, dy/dt \approx -dv_0/d\sigma$ for $t > 0$. Thus, $\mu_0(\tau)$ must be a decaying solution of

$$\frac{d\mu_0}{d\tau} = g(X_0(0), Y_0(0) + \mu_0(\tau), 0, 0), \qquad \mu_0(0) = y(0,0) - Y_0(0)$$

on $\tau \ge 0$, whereas $v_0(\sigma)$ must be a decaying solution of

$$\frac{dv_0}{d\sigma} = -g(X_0(1), Y_0(1) + v_0(\sigma), 1, 0), \qquad v_0(0) = y(1,0) - Y_0(1).$$

The differential systems for the limiting solution and the limiting boundary layer terms are coupled through the boundary condition

$$a(X_0(0), Y_0(0) + \mu_0(0), X_0(1), Y_0(1) + v_0(0), 0) = 0.$$

That we can actually often obtain isolated solutions of the combined problem follows under many hypothesis; for example, when (i) the system is linear in the fast variables, (ii) the Jacobian matrix g_y is nonsingular with a fixed eigenvalue splitting, and (iii) appropriate boundary conditions are prescribed. More interesting and challenging examples occur when interior shock or transition layers of nonuniform convergence occur within (0,1). In contrast, of course, we also have examples like that of Coddington and Levinson (1952) (see Section 1D) where no solution exists for very small ϵ's. Some quite sophisticated and promising geometric approaches to such problems are contained in Fenichel (1979), Kurland (1986), Pollack (1989), Lin (1989), Szmolyan (1989), and Kopell (1990) among other papers. New perspectives are still needed.

K. A Nonlinear Control Problem

Suppose we want to solve the singularly perturbed initial value problem

$$\begin{cases} \dot{x} = f(x, z, u, t, \epsilon), & x(0) \text{ given}, \\ \epsilon \dot{z} = g(x, z, u, t, \epsilon), & z(0) \text{ given} \end{cases}$$

consisting of $m+n$ differential equations on the interval $0 \leq t \leq 1$, where the optimal control vector $u(t)$ is selected to minimize the scalar cost functional

$$J(u) = \int_0^1 \Lambda(x(t), z(t), u(t), t, \epsilon) dt.$$

Such nonlinear regulator problems, especially various special cases, are surveyed in O'Malley (1978), Kokotovic (1984), Kokotovic et al. (1986), Bensoussan (1988), and Kushner (1990). Necessary conditions for optimality follow from the Pontryagin maximum principle and other variational arguments [cf., e.g., Fleming and Rishel (1975)]. If we introduce the Hamiltonian

$$h = \Lambda + p'f + q'g,$$

where the costate vectors (i.e., Lagrange multipliers) p and q satisfy the terminal value problem

$$\begin{cases} \dot{p} = -h_x, & p(1) = 0, \\ \epsilon \dot{q} = -h_z, & q(1) = 0, \end{cases}$$

the state equations can be written as

$$\begin{cases} \dot{x} = h_p, & x(0) \text{ given}, \\ \epsilon \dot{z} = h_q, & z(0) \text{ given} \end{cases}$$

and the optimality condition becomes

$$h_u = 0.$$

Since we naturally want to minimize h, the Legendre–Clebsch condition $h_{uu} \geq 0$ must also hold. In its strengthened form, when the Hessian matrix h_{uu} is positive definite (at least locally), we can solve $h_u = 0$ to obtain the optimal control

$$u = \xi(x, z, p, q, t, \epsilon).$$

Substituting this u back into the state and costate equations then reduces our optimal control problem to solving a nonlinear singularly perturbed two-point boundary value problem. That problem is quasilinear, and quite tractable, when f and g are linear and the Lagrangian Λ is quadratic in z and u. The situation becomes more complicated and more challenging when the control u is allowed to switch abruptly within $[0,1]$ from one root of $h_u = 0$ to another. This, of course, requires h to be at least quadratic in u. Layer phenomena must certainly be expected at such switching points.

The boundary value problem so obtained always has a special *symplectic* structure which should be maintained in determining its asymptotic solution. This becomes more obvious when we introduce the slow and fast vectors

$$\psi = \begin{pmatrix} x \\ p \end{pmatrix}, \qquad \zeta = \begin{pmatrix} z \\ q \end{pmatrix}$$

and rewrite the Hamiltonian as

$$H(\psi, \zeta, u, t, \epsilon) = h(x, z, u, p, q, t, \epsilon).$$

The state and costate equations can then be rewritten as

$$\dot{\psi} = J_m H_\psi, \qquad \epsilon \dot{\zeta} = J_n H_\zeta,$$

where J_k denotes the symplectic matrix $J_k = \begin{pmatrix} 0 & I_k \\ -I_k & 0 \end{pmatrix}$. If the Hessian $H_{uu} = h_{uu}$ is nonsingular, and we solve the optimality condition $H_u = 0$ for the optimal

$$u = \eta(\psi, \zeta, t, \epsilon) = \xi(x, z, p, q, t, \epsilon),$$

we obtain the nonlinear singularly perturbed $2(m+n) \times 2(m+n)$-dimensional system

$$\dot{\psi} = F(\psi, \zeta, t, \epsilon) \equiv J_m H_\psi(\psi, \zeta, \eta, t, \epsilon),$$

$$\epsilon \dot{\zeta} = G(\psi, \zeta, t, \epsilon) \equiv J_n H_\zeta(\psi, \zeta, \eta, t, \epsilon).$$

It is subject to the $2(m+n)$ inherited boundary conditions

$$\begin{pmatrix} I_m & 0 \\ 0 & 0 \end{pmatrix} \psi(0) = \begin{pmatrix} x(0) \\ 0 \end{pmatrix} \quad \text{and} \quad \begin{pmatrix} I_n & 0 \\ 0 & 0 \end{pmatrix} \zeta(0) = \begin{pmatrix} z(0) \\ 0 \end{pmatrix}$$

and

$$\begin{pmatrix} 0 & 0 \\ 0 & I_m \end{pmatrix} \psi(1) = \begin{pmatrix} 0 \\ p(1) \end{pmatrix} = 0 \quad \text{and} \quad \begin{pmatrix} 0 & 0 \\ 0 & I_n \end{pmatrix} \zeta(1) = \begin{pmatrix} 0 \\ q(1) \end{pmatrix} = 0.$$

It is most important to observe that the critical $2n \times 2n$ Jacobian G_ζ is a Hamiltonian matrix. A simple calculation shows that

$$G_\zeta = J_n \left(H_{\zeta\zeta} + H_{\zeta u} \frac{\partial \eta}{\partial \zeta} \right) = J_n (H_{\zeta\zeta} - H_{u\zeta}' H_{uu}^{-1} H_{u\zeta})$$

since $H_u \equiv 0$ implies that $H_{uu}(\partial \eta / \partial \zeta) + H_{u\zeta} = 0$, whereas $H_{\zeta u} = H_{u\zeta}'$. Thus, $J_n G_\zeta$ is symmetric. Moreover, this implies that the eigenvalues of G_ζ occur in pairs $\pm\lambda$ and the corresponding eigenvectors are likewise specially related [cf. Coppel (1975)].

If we now *assume* that (locally) G_ζ has no purely imaginary eigenvalues (thereby eliminating the possibility of turning points as well as highly oscillatory solutions), G_ζ will have n strictly stable and n strictly unstable eigenvalues. Then, any limiting solution can be expected to satisfy the reduced (differential-algebraic) system

$$\begin{cases} \dot{\Psi}_0 = F(\Psi_0, Z_0, t, 0), \\ 0 = G(\Psi_0, Z_0, t, 0). \end{cases}$$

Moreover, the nonsingularity of G_ζ implies that the algebraic system $G = 0$ can be solved (at least locally) for some

$$Z_0 = \phi(\Psi_0, t),$$

leaving us the $2m$th-order system

$$\dot{\Psi}_0 = \mathcal{F}(\Psi_0, t) \equiv F(\Psi_0, \phi(\Psi_0(t), t), t, 0).$$

This reduction makes it more natural to prescribe boundary values for Ψ_0, rather than Z_0. If we retain the boundary condition prescribing $x(0)$ as well as $p(1) = 0$, the resulting reduced problem

$$
\begin{cases}
\dot{\Psi}_0 = \mathcal{F}(\Psi_0, t), \\
\begin{pmatrix} I_m & 0 \\ 0 & 0 \end{pmatrix} \Psi_0(0) = \begin{pmatrix} x(0) \\ 0 \end{pmatrix}, \quad \begin{pmatrix} 0 & 0 \\ 0 & I_m \end{pmatrix} \Psi_0(1) = \begin{pmatrix} 0 \\ p(1) \end{pmatrix} = 0
\end{cases}
$$

will inherit the Hamiltonian structure through our sequence of transformations. Indeed, under the accumulated assumptions, we could show that the reduced problem provides a solution to the reduced optimal control problem, i.e., it solves the initial value problem

$$
\begin{cases}
\dot{x} = f(x, z, U, t, 0) \text{ with } x(0) \text{ given} \\[1ex]
\text{subject to the constraint} \\[1ex]
g(x, z, U, t, 0) = 0 \\[1ex]
\text{where } U \text{ minimizes the cost functional} \\[1ex]
J(U) = \int_0^1 \Lambda(x, z, U, t, 0) dt.
\end{cases}
$$

Because the initial condition for z and the terminal condition for q will not generally be satisfied by the solution of the reduced optimal control problem, the fast variable ζ will require both an initial and a terminal layer correction. Thus, we will seek ζ in the form

$$\zeta(t) = Z_0(t) + \zeta_{L0}(\tau) + \zeta_{R0}(\sigma) + O(\epsilon),$$

where ζ_{L0} (and ζ_{R0}) $\to 0$ as $\tau = t/\epsilon$ [and $\sigma = (1-t)/\epsilon$, respectively] $\to \infty$. As usual, ζ_{L0} will be a decaying solution of the $2n$-dimensional conditionally stable initial layer system

$$
\begin{cases}
\dfrac{d\zeta_{L0}}{d\tau} = G(\Psi_0(0), Z_0(0) + \zeta_{L0}, 0, 0) \\[2ex]
\text{with } \begin{pmatrix} I_n & 0 \\ 0 & 0 \end{pmatrix} (Z_0(0) + \zeta_{L0}(0)) = \begin{pmatrix} z(0) \\ 0 \end{pmatrix} \text{ prescribed.}
\end{cases}
$$

In analogous fashion, ζ_{R0} must be a decaying solution of the terminal layer system

$$\begin{cases} \dfrac{d\zeta_{R0}}{d\sigma} = -G(\Psi_0(1), Z_0(1) + \zeta_{R0}, 1, 0) \\[2mm] \text{with } \begin{pmatrix} 0 & 0 \\ 0 & I_n \end{pmatrix}(Z_0(1) + \zeta_{R0}(0)) = 0. \end{cases}$$

These two layer problems will be well posed provided the initial vectors lie on the n-dimensional stable initial manifolds for the respective systems on the intervals $\tau \geq 0$ or $\sigma \geq 0$. In the special case that the stable manifold for $\zeta_{L0} = \binom{z_{L0}}{q_{L0}}$ can be described in the form

$$q_{L0} = \Phi^L(z_{L0}),$$

we can replace our $2n$th-order conditionally stable limiting initial layer system with a stable nth-order initial value problem

$$\frac{dz_{L0}}{d\tau} = g_L(z_{L0}) \equiv G^1(\Psi_0(0), \qquad Z_0(0) + \begin{pmatrix} z_{L0} \\ \Phi^L(z_{L0}) \end{pmatrix}, 0, 0),$$

with $z_{L0}(0)$ prescribed.

Here, G^1 consists of the first n rows of G and, under natural conditions, $\partial G^1/\partial z_{L0}$ will have the n stable eigenvalues of G_ζ [cf. Gu (1987)]. Under analogous assumptions, a similar reduction will be possible for the limiting terminal layer problem. These two limiting layer problems can each also be further interpreted as providing the solution of certain natural infinite-interval regulator problems, which likewise naturally possess a Hamiltonian structure. Thus, corresponding to the initial layer problem, we might associate a nonlinear infinite interval regulator problem, and seek a decaying solution of the initial value problem

$$\frac{dz}{d\tau} = g(x(0), z, w_L, 0, 0) \text{ for an appropriate } z(0) \text{ on } \tau \geq 0,$$

where the control w_L should minimize its cost

$$J_L(w_L) = \int_0^\infty [\Lambda(x(0), z(s), w_L(s), 0, 0) - \Lambda(x(0), 0, 0, 0, 0)]ds.$$

Under appropriate hypothesis, we can also expect the optimal control $u(t)$ for our original regulator problem to have an asymptotic limit in the composite form

$$u(t) = U(t) + w_L(\tau) + w_R(\sigma) + O(\epsilon),$$

where U, w_L, and w_R are, respectively, the optimal controls for the reduced optimal control problem, for the initial layer control problem, and for the terminal layer control problem. [Another valuable approach to obtaining such a composite control is given in Khalil and Hu (1989)]. We can get higher-order approximations to the states, costates, and optimal control by successively solving corresponding linearized problems. For them, the theory is quite thoroughly developed [cf. Kokotovic (1984)]. We note that more details concerning this approach are contained in O'Malley (1978), based on unfinished work of McIntyre [cf. also the somewhat related work of Anderson and Kokotovic (1987)].

In summary, we note that this nonlinear control problem is intermediate in difficulty between general nonlinear two-point problems and the well-understood quasilinear problems. The special Hamiltonian structure is of independent interest. Moreover, the asymptotic solutions obtained are of practical value in applications.

Example

Consider the scalar problem where the state z satisfies the nonlinear singularly perturbed equation

$$\epsilon\dot{z} = z^3 + u + 1, \qquad z(0) = 1,$$

and the optimal control u minimizes the cost

$$J(u) = \frac{1}{2}\int_0^1 (z^2 + u^2)dt.$$

Such problems were treated in Kokotovic et al. (1986) using Bellman's principle of optimality. Let us, instead, introduce the Hamiltonian

$$h = \tfrac{1}{2}(z^2 + u^2) + q(z^3 + u + 1),$$

where the costate q satisfies

$$\epsilon\dot{q} = -h_z = -z - 3qz^2, \qquad q(1) = 0.$$

Note that the state and costate equations form the singularly perturbed Hamilton–Jacobi system $\epsilon\dot{z} = h_q, \epsilon\dot{q} = -h_z$. The optimality condition $h_u = 0$ implies that

$$u + q = 0,$$

so the necessary conditions for optimality reduce to solving the nonlinear singularly perturbed boundary value problem

$$\begin{cases} \epsilon\dot{z} = z^3 - q + 1, & z(0) = 1, \\ \epsilon\dot{q} = -z - 3qz^2, & q(1) = 0. \end{cases}$$

This strongly nonlinear problem seems beyond the scope of our known theory. The corresponding reduced problem is simply the algebraic system

$$\begin{cases} Z_0^3 - Q_0 + 1 = 0, \\ -Z_0 - 3Q_0 Z_0^2 = 0. \end{cases}$$

Thus,

$$Q_0 = 1 + Z_0^3$$

and Z_0 must satisfy the quintic polynomial

$$Z_0 + 3Z_0^2 + 3Z_0^5 = 0.$$

The three real solutions are (approximately) given by $Z_0 = -0.846, -0.348$, and 0. Since the corresponding costs of these outer solutions [i.e., $\int_0^1 \frac{1}{2}(Z_0^2 + Q_0^2)dt$] are, respectively, given by $0.436, 0.519$, and 0.500, and any boundary layer corrections will only contribute asymptotically small amounts to the full cost, we take as optimal

$$Z_0 = -0.846 \quad \text{and} \quad Q_0 = 0.394.$$

Because this outer solution does not satisfy either boundary condition, there is need for a boundary layer at each endpoint. [We note that this selection corresponds, also, to having a conditionally stable Jacobian (evaluated along Z_0 and Q_0).] Thus, we look for a solution of the form

$$z = z_L(\tau) + Z_0 + z_R(\sigma),$$

$$q = q_L(\tau) + Q_0 + q_R(\sigma),$$

where z_L and $q_L \to 0$ as $\tau = t/\epsilon \to \infty$ and z_R and $q_R \to 0$ as $\sigma = (1-t)/\epsilon \to \infty$. Since the terminal layer correction is negligible for $t < 1$, the initial layer correction $\binom{z_L}{q_L}$ must satisfy the stretched problem

$$\frac{dz_L}{d\tau} = [(z_L + Z_0)^3 - (q_L + Q_0) + 1] - (Z_0^3 - Q_0 + 1),$$

$$\frac{dq_L}{d\tau} = -(z_L + Z_0) - 3(z_L + Z_0)^2(q_L + Q_0) - (-Z_0 - 3Z_0^2 Q_0).$$

Thus, it must provide decaying solutions of the initial value problem

$$\frac{dz_L}{d\tau} = z_L^3 + 3z_L^2 Z_0 + 3z_L Z_0^2 - q_L, \qquad z_L(0) = 1 - Z_0,$$

$$\frac{dq_L}{d\tau} = -z_L - 3(z_L^2 + 2z_L Z_0 + Z_0^2)q_L - 3(z_L^2 + 2z_L Z_0)Q_0.$$

Eliminating q_L, these equations imply the scalar conservative equation

$$\frac{d^2 z_L}{d\tau^2} = z_L(1 + 6Z_0 + 15Z_0^4) + 3z_L^2(1 + 10Z_0^3) + 30z_L^3 Z_0^2 + 15z_L^4 Z_0 + 3z_L^5.$$

Multiplying by $2(dz_L/d\tau)$ and integrating implies that

$$\left(\frac{dz_L}{d\tau}\right)^2 = z_L^2\big[(1 + 6Z_0 + 15Z_0^4) + 2z_L(1 + 10Z_0^3)$$

$$+ 15z_L^2 Z_0^2 + 6z_L^3 Z_0 + z_L^4\big]$$

$$\equiv z_L^2 g^2(z_L) \qquad \text{with } g(z_L) \geq 0.$$

Since $z_L(0) = 1 - Z_0 > 0$, our decay requirement implies that we must take

$$\frac{dz_L}{d\tau} = -z_L g(z_L).$$

Since $g(z_L)$ has a positive minimum at about 0.959, $dz_L/d\tau$ has no zeros between $z_L(0)$ and 0. Thus, z_L will decay monotonically and, ultimately, exponentially toward zero as $\tau \to \infty$. In terms of z_L, which we could readily obtain numerically, we have

$$q_L(\tau) = z_L g(z_L) + 3z_L Z_0^2 + 3z_L^2 Z_0 + z_L^3.$$

It, too, decreases monotonically to zero.

The terminal layer problem is equivalent to finding a decaying solution of the initial value problem

$$\begin{cases} \dfrac{dz_R}{d\sigma} = q_R - 3Z_0^2 z_R - 3Z_0 z_R^2 - z_R^3, \\[2mm] \dfrac{dq_R}{d\sigma} = z_R + 3q_R(z_R^2 + 2z_R Z_0 + Z_0^2) + 3Q_0(z_R^2 + 2z_R Z_0), \quad q_R(0) = -Q_0. \end{cases}$$

Eliminating q_R, we get a conservative equation for z_R. Integration now yields

$$\frac{dz_R}{d\sigma} = \pm z_R g(z_R).$$

Because $q_R + Z_0^3 - (Z_0 + z_R)^3 = dz_R/d\sigma$, the initial condition $q_R(0) = -Q_0 = -(1 + Z_0^3)$ implies that $z_R(0)$ must satisfy the quadratic equation

$$z_R^2(0) - 6Z_0^2(1 + Z_0^3)z_R(0) - (1 + Z_0^3)^2 = 0,$$

so

$$z_R(0) = (1 + Z_0^3)\left(3Z_0^2 \pm \sqrt{1 + 9Z_0^4}\right).$$

Numerically, the possible $z_R(0)$'s are approximately equal to 1.78 or -0.087. The corresponding solutions $z_R(\sigma)$ will decay monotonically to zero as $\sigma \to \infty$ provided we pick the sign of $dz_R/d\sigma$ opposite that of z_R, since $g(z_R)$ remains positive. The appropriate choice of $z_R(0)$ results from minimizing the cost of the terminal layer correction, i.e., the integral

$$\frac{1}{2}\int_0^\infty \{[Z_0 + z_R(\sigma)]^2 + [Q_0 + q_R(\sigma)]^2 - (Z_0^2 + Q_0^2)\}d\sigma.$$

Thus, our nonlinear control problem has a unique and relatively simple solution. Readers should consider how general a class of nonlinear boundary value problems could be successfully treated in a similar fashion.

L. Semiconductor Modeling

Semiconductor device modeling continues to provide an important and challenging application of asymptotic methods for singularly perturbed boundary value problems [cf., e.g., Markowich (1986) and Markowich et al. (1990)]. Vasil'eva and her Soviet colleagues, among others throughout the world, have completed extensive investigations in the field [cf. Vasil'eva et al. (1976)]. The relative importance of different layer phenomena, especially from the perspective of simulation and engineering design, was emphasized by Selberherr (1988).

Vasil'eva and Stelmakh (1977) and Schmeiser and Weiss (1986) have considered the symmetric p–n junction with piecewise constant doping. With an appropriate scaling, the resulting one-dimensional stationary problem is given by

$$\begin{cases} \epsilon\dot{e} = n - p - 1, \\ \\ \epsilon\dot{p} = -pe - \dfrac{\epsilon}{2}J, \quad p(1) = \dfrac{\delta^2}{n(1)}, \\ \\ \epsilon\dot{n} = ne + \dfrac{\epsilon}{2}J, \quad n(0) = p(0), \ n(1) = p(1) + 1, \\ \\ \epsilon\dot{\psi} = e, \qquad \psi(0) = 0 \end{cases}$$

on $0 \leq x \leq 1$ where the small positive parameter ϵ represents a scaled Debye length (typically about $10^{-7/2}$) and where J and δ are fixed positive constants representing the current density and an intrinsic carrier concentration. Employing symmetry, $x = 0$ corresponds to the location of the p–n junction (in the middle of the physical device) and $x = 1$ is the location of one ohmic contact. Solving for the potential

$$\psi(x) = \frac{1}{\epsilon}\int_0^x e(s)ds,$$

there remains a third-order singularly perturbed system for the electric field e, the (non-negative) hole density p, and the (non-negative) electron density n. The sign of p determines the terminal values as

$$p(1) = \tfrac{1}{2}\left(-1 + \sqrt{1 + 4\delta^2}\right)$$

and
$$n(1) = p(1) + 1.$$
The corresponding limiting outer (or reduced) problem
$$\begin{cases} N_0 - P_0 - 1 = 0, \\ -P_0 E_0 = 0, \\ N_0 E_0 = 0 \end{cases}$$
determines the trivial limiting field $E_0 = 0$ and demands a trivial limiting space charge (i.e., $N_0 - P_0 = 1$, consistent with the terminal condition), but leaves P_0 otherwise unspecified. This is, of course, typical of singular singular-perturbation problems and results because the Jacobian of the system, evaluated along any such limiting solution, is singular. (The Jacobian has, indeed, one trivial, one positive, and one negative eigenvalue.) Since the limiting outer solution's zero space charge contradicts the initial condition $n(0) = p(0)$, we must expect a region of nonuniform convergence as $\epsilon \to 0$ near the p–n junction at $x = 0$.

Let us introduce
$$w = np,$$
noting that $\dot{w}$ will remain bounded as $\epsilon \to 0$. Eliminating the fast variable $n = w/p$ in terms of the slow w, we convert our singular singular-perturbation problem into the traditional fast–slow form
$$\begin{cases} \epsilon \dot{e} = \dfrac{w}{p} - p - 1, \\[2mm] \epsilon \dot{p} = -pe - \dfrac{\epsilon}{2} J, \quad p^2(0) = w(0), \quad p(1) = \dfrac{1}{2}\left(-1 + \sqrt{1 + 4\delta^2}\right), \\[2mm] \dot{w} = \dfrac{J}{2}\left(p - \dfrac{w}{p}\right), \qquad w(1) = \delta^2. \end{cases}$$

Here, we naturally take the reduced problem to be
$$\begin{cases} 0 = \dfrac{W_0}{P_0} - P_0 - 1, \\[2mm] 0 = -P_0 E_0, \\[2mm] \dot{W}_0 = \dfrac{J}{2}\left(P_0 - \dfrac{W_0}{P_0}\right), \qquad W_0(1) = \delta^2, \end{cases}$$
since the conditional stability of the fast system suggests cancelling one initial and one terminal condition, while a boundary condition is needed to specify the limiting slow variable W_0. Since the outer solution satisfies $P_0^2 + P_0 = W_0$, while $\dot{W}_0 = -J/2$, we must have

$$
\begin{cases}
W_0(x) = \delta^2 + \dfrac{J}{2}(1 - x), \\[2mm]
E_0(x) = 0, \\[2mm]
P_0(x) = \tfrac{1}{2}\left[-1 + \sqrt{1 + 4W_0(x)}\right]
\end{cases}
$$

because $P_0(x) \geq 0$. For w to converge uniformly to $W_0(0)$ at $x = 0$, we must take

$$
p(0) = \sqrt{\delta^2 + J/2}.
$$

Then, $P_0(0) \neq p(0)$ (in general), so we must anticipate that the limiting behavior at $x = 0$ could be described in terms of the limiting conditionally stable inner problem

$$
\begin{cases}
\dfrac{d\tilde{e}}{d\tau} = \dfrac{p^2(0)}{\tilde{p}} - \tilde{p} - 1, \\[3mm]
\dfrac{d\tilde{p}}{d\tau} = -\tilde{p}\tilde{e}, \qquad \tilde{p}(0) = p(0)
\end{cases}
$$

on $\tau = x/\epsilon \geq 0$, subject to the matching condition

$$
(\tilde{e}(\infty), \tilde{p}(\infty)) = (E_0(0), P_0(0)) = \left(0, \tfrac{1}{2}\left[-1 + \sqrt{1 + 4p^2(0)}\right]\right).
$$

In the $\tilde{e}$–$\tilde{p}$ phase plane, we will have

$$
\tilde{e}\dfrac{d\tilde{e}}{d\tilde{p}} = -\dfrac{p^2(0)}{\tilde{p}^2} + 1 + \dfrac{1}{\tilde{p}}.
$$

Thus, integration backward from infinity implies that

$$
\dfrac{1}{2}\tilde{e}^2 = \dfrac{p^2(0)}{\tilde{p}} + \tilde{p} + \ln\dfrac{\tilde{p}}{P_0(0)} + \sqrt{1 + 4p^2(0)}.
$$

This defines the stable manifold $\tilde{e}$ as a function of $\tilde{p}$ once we select the sign of $\tilde{e}$ to guarantee that the remaining equation

$$
\dfrac{1}{\tilde{p}}\dfrac{d\tilde{p}}{d\tau} = -\tilde{e}
$$

has a solution $\tilde{p}(\tau)$ which changes monotonically from $p(0)$ to $\tilde{p}(\infty) = P_0(0)$ as τ increases. Note that the initial value $e(0)$ of the electric field is determined by $p(0)$ and that a numerical integration of the nonlinear initial value problem for $\tilde{p}(\tau)$ on $\tau \geq 0$ should be relatively straightforward.

The local behavior of the solution near $x = 1$ will be determined by the limiting inner problem

$$\begin{cases} \dfrac{d\hat{e}}{d\sigma} = -\dfrac{\delta^2}{\hat{p}} + \hat{p} + 1, \\[3mm] \dfrac{d\hat{p}}{d\sigma} = \hat{p}\hat{e}, \qquad \hat{p}(0) = p(1) \end{cases}$$

on the interval $\sigma = (1 - x)/\epsilon \geq 0$ and the matching condition

$$(\hat{e}(\infty), \hat{p}(\infty)) = (E_0(1), P_0(1)) = (0, p(1)).$$

It is straightforward to show that this conditional stability problem only has the constant solution $(\hat{e}(\sigma), \hat{p}(\sigma))$. The resulting absence of a terminal layer corresponds to uniform convergence of the solution at the ohmic contact.

Higher-order matching at both endpoints could be used to obtain uniform asymptotic approximations to the solution. Alternatively, in terms of the original variables, it is possible to proceed directly using the ansatz

$$\begin{cases} \psi(t, \epsilon) = \Psi(t, \epsilon) + \alpha(\tau, \epsilon) + \epsilon\mu(\sigma, \epsilon), \\[2mm] e(t, \epsilon) = \epsilon\dot{\Psi}(t, \epsilon) + \dfrac{d\alpha}{d\tau}(\tau, \epsilon) - \epsilon\dfrac{d\mu}{d\sigma}(\sigma, \epsilon), \\[2mm] p(t, \epsilon) = P(t, \epsilon) + \beta(\tau, \epsilon) + \epsilon\nu(\sigma, \epsilon), \\[2mm] n(t, \epsilon) = N(t, \epsilon) + \gamma(\tau, \epsilon) + \epsilon\rho(\sigma, \epsilon), \end{cases}$$

where all terms have an asymptotic power series expansion in ϵ, where the functions of $\tau = x/\epsilon$ decay to zero as $\tau \to \infty$, and where the functions of $\sigma = (1 - x)/\epsilon$ decay to zero as $\sigma \to \infty$. The solution, then, exhibits nonuniform convergence at the junction point $x = 0$ (due to the jump in the doping), while its derivative also does so at the contact at $x = 1$. Knowing this asymptotic structure of the solution is, of course, extremely valuable (perhaps critical) in designing numerical algorithms to obtain solutions.

M. Shocks and Transition Layers

Let us now consider the scalar two-point problem

$$\begin{cases} \epsilon\ddot{x} + f(x, t)\dot{x} + g(x, t) = 0, \qquad 0 \leq t \leq 1, \\[2mm] x(0) = A, \quad x(1) = B \end{cases}$$

and seek a solution which has a transition layer of nonuniform convergence at some interior point $\tilde{t}$ in $0 < t < 1$ and which converges uniformly elsewhere. Pictorially, we are seeking solutions of the form shown in Figure 3.7. We note that some vector versions of such problems have been considered by Howes and O'Malley (1980), Chang and Howes (1984), and Kelley (1988), while computational aspects are considered in Hedstrom and Howes (1990).

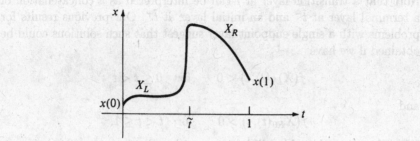

Fig. 3.7. An interior shock.

The asymptotic solutions for $t < \tilde{t}$ and $t > \tilde{t}$ can then be found by regular perturbation methods. Specifically, let $X_L(t, \epsilon)$ and $X_R(t, \epsilon)$ represent those asymptotic solutions where X_L has an outer expansion

$$X_L(t, \epsilon) \sim \sum_{j=0}^{\infty} X_{Lj}(t)\epsilon^j \qquad \text{on } 0 \leq t \leq \tilde{t}.$$

Necessarily, the limit X_{L0} must satisfy the reduced problem

$$f(X_{L0}, t)\dot{X}_{L0} + g(X_{L0}, t) = 0, \qquad X_{L0}(0) = A,$$

whereas later terms X_{Lj} for $j > 0$ will successively satisfy linearized problems

$$f(X_{L0}, t)\dot{X}_{Lj} + f_x(X_{L0}, t)X_{Lj}\dot{X}_{L0} + g_x(X_{L0}, t)X_{Lj} = \alpha_{j-1}(t), X_{Lj}(0) = 0,$$

where α_{j-1} is specified in terms of preceding coefficients. We note that these problems will all be uniquely solvable on $0 \leq t \leq \tilde{t}$ if the reduced problem has a solution $X_{L0}(t)$ which exists throughout the interval and along which

$$f(X_{L0}(t), t) \neq 0.$$

In analogous fashion, we can determine an outer expansion

$$X_R(t, \epsilon) \sim \sum_{j=0}^{\infty} X_{Rj}(t)\epsilon^j \qquad \text{on } \tilde{t} \leq t \leq 1$$

satisfying the terminal condition, provided the reduced problem

$$f(X_{R0}, t)\dot{X}_{R0} + g(X_{R0}, t) = 0, \qquad X_{R0}(1) = B$$

has a solution $X_{R0}(t)$ there on which it maintains

$$f(X_{R0}, t) \neq 0.$$

Note that a transition layer at $\tilde{t}$ can be interpreted as a concatenation of a terminal layer at $\tilde{t}^-$ and an initial layer at $\tilde{t}^+$. Our previous results for problems with a single endpoint layer suggest that such solutions could be obtained if we have

$$f(X_{L0}(t), t) < 0 \qquad \text{for} \quad 0 \leq t < \tilde{t}$$

and

$$f(X_{R0}(t), t) > 0 \qquad \text{for} \quad \tilde{t} < t \leq 1.$$

We will *assume* that this will be so, even though $\tilde{t}$ is not yet fixed.

To analyze the transition near $\tilde{t}$, we introduce the stretched variable

$$\kappa = (t - \tilde{t})/\epsilon$$

and seek a local asymptotic solution of the form

$$x(t, \epsilon) = X_L(t, \epsilon) + \mu(\kappa, \epsilon)$$

where the transition layer correction μ satisfies

$$\begin{cases} \mu \to 0 & \text{as } \kappa \to -\infty, \\ \mu \to X_R(\tilde{t} + \epsilon\kappa, \epsilon) - X_L(\tilde{t} + \epsilon\kappa, \epsilon) & \text{as } \kappa \to +\infty. \end{cases}$$

This approach, somewhat implicitly, requires there to be an appropriate overlapping of the domains of definition of the left and right outer expansions X_L and X_R. In particular, by using their Taylor expansions about $\epsilon = 0$, we will obtain the expansion

$$\mu(\kappa, \epsilon) \sim X_R(\tilde{t} + \epsilon\kappa, \epsilon) - X_L(\tilde{t} + \epsilon\kappa, \epsilon) \sim \sum_{j=0}^{\infty} \mu_j(\kappa)\epsilon^j$$

as $\kappa \to \infty$, where the μ_j's are polynomials in κ of degree j or less. [An alternative tack would be to seek a transition layer expansion $\phi(\kappa, \epsilon) \sim \sum_{j=0}^{\infty} \phi_j(\kappa)\epsilon^j$ which matches the outer expansions $X_R(t, \epsilon)$ and $X_L(t, \epsilon)$ as $\kappa \to \pm\infty$, respectively.] Here, the transition layer correction μ will necessarily satisfy the nonlinear equation

$$\frac{d^2\mu}{d\kappa^2} + f(X_L(\tilde{t} + \epsilon\kappa, \epsilon) + \mu, \tilde{t} + \epsilon\kappa)\frac{d\mu}{d\kappa}$$

$$+ \epsilon[f(X_L(\tilde{t} + \epsilon\kappa, \epsilon) + \mu, \tilde{t} + \epsilon\kappa)$$

$$- f(X_L(\tilde{t} + \epsilon\kappa, \epsilon), \tilde{t} + \epsilon\kappa)]\frac{dX_L}{dt}(\tilde{t} + \epsilon\kappa, \epsilon)$$

$$+ \epsilon[g(X_L(\tilde{t} + \epsilon\kappa, \epsilon) + \mu, \tilde{t} + \epsilon\kappa) - g(X_L(\tilde{t} + \epsilon\kappa, \epsilon), \tilde{t} + \epsilon\kappa)] = 0$$

on $-\infty < \kappa < \infty$ as well as the boundary conditions. The leading term μ_0 must therefore satisfy

$$\frac{d^2\mu_0}{d\kappa^2} + f(X_{L0}(\tilde{t}) + \mu_0, \tilde{t})\frac{d\mu_0}{d\kappa} = 0.$$

Since μ_0 must tend to zero as $\kappa \to -\infty$, integration provides $d\mu_0/d\kappa$.

There remains the terminal value problem

$$\frac{d\mu_0}{d\kappa} + \int_0^{\mu_0(\kappa)} f(X_{L0}(\tilde{t}) + r, \tilde{t})dr = 0, \quad \mu_0(\infty) = X_{R0}(\tilde{t}) - X_{L0}(\tilde{t})$$

to be solved on $-\infty < \kappa < \infty$. The solution is, of course, trivial if $X_{L0}(\tilde{t}) = X_{R0}(\tilde{t})$ (since there is then no need for a transition layer). Otherwise, we would obtain a monotonic solution provided

$$(*) \qquad \mu_0(\infty) \int_{X_{L0}(\tilde{t})}^{v} f(s,\tilde{t})ds < 0$$

for all v *between* $X_{L0}(\tilde{t})$ and $X_{R0}(\tilde{t})$. Note that this hypothesis requires $f(X_{L0}(\tilde{t}), \tilde{t}) < 0$, corresponding to the already imposed stability requirement on $X_{L0}(t)$ for $t < \tilde{t}$, and likewise that $f(X_{R0}(t), t) > 0$ for $t > \tilde{t}$. Moreover, to have a rest point at $\kappa = \infty$, we will need

$$(**) \qquad S(\tilde{t}) \equiv \int_{X_{L0}(\tilde{t})}^{X_{R0}(\tilde{t})} f(s,\tilde{t})ds = 0.$$

The conditions $(*)$ and $(**)$ are convenient and natural, but *not necessary*, for obtaining a transition layer. Such conditions correspond to the well-known entropy and Rankine–Hugoniot conditions of gas dynamics [cf. Smoller (1983)]. We note that the condition $S(\tilde{t}) = 0$ will determine the location $\tilde{t}$ of the jump (at least locally) provided $\dot{S}(\tilde{t}) \neq 0$ or $\tilde{t}$ is otherwise known to be an isolated zero of S.

Example

For the Lagerstrom–Cole example

$$\epsilon\ddot{x} + x\dot{x} - x = 0, \qquad x(0) = A, x(1) = B,$$

the reduced problems

$$X_{L0}\dot{X}_{L0} - X_{L0} = 0, \qquad X_{L0}(0) = A, 0 \le t < \tilde{t},$$

and

$$X_{R0}\dot{X}_{R0} - X_{R0} = 0, \qquad X_{R0}(1) = B, \tilde{t} < t \le 1$$

will have stable solutions

$$X_{L0}(t) = t + A$$

and

$$X_{R0}(t) = t + B - 1,$$

on the respective intervals, provided

$$\tilde{t} + A < 0 \qquad \text{and} \qquad \tilde{t} + B - 1 > 0$$

[since $f(X_{L0}, t) < 0$ and $f(X_{R0}, t) > 0$ will then hold]. The resulting transition layer problem

$$\frac{d\mu_0}{d\kappa} = -\int_0^{\mu_0} [X_{L0}(\tilde{t}) + r] dr = \tfrac{1}{2}\{X_{L0}^2(\tilde{t}) - [X_{L0}(\tilde{t}) + \mu_0]^2\},$$

with $\mu_0(\infty) = B - A - 1$, can be integrated (as a Riccati equation) in terms of a hyperbolic tangent function. To obtain a rest point at infinity requires

$$X_{L0}^2(\tilde{t}) = X_{R0}^2(\tilde{t}),$$

so to obtain a jump, we must take

$$X_{L0}(\tilde{t}) = -X_{R0}(\tilde{t}) < 0.$$

This uniquely specifies the jump location to be at

$$\tilde{t} = \tfrac{1}{2}(1 - A - B).$$

Moreover, the entropy condition will also be satisfied. To have $0 < \tilde{t} < 1$, we must have

$$-1 < A + B < 1.$$

Since stability requires $X_{L0}(\tilde{t})$ to be negative, we also need

$$B > A + 1.$$

For other values of A and B, it follows that the asymptotic solution cannot have such an interior transition layer [cf. Kevorkian and Cole (1981)]. The solution can then be shown to feature endpoint layers and/or angular solutions which involve the trivial limiting solution.

Returning to the general problem, note that higher-order terms μ_j in the expansion for a transition layer correction must satisfy a linear differential equation

$$\frac{d^2\mu_j}{d\kappa^2} + f(X_{L0}(\tilde{t}) + \mu_0, \tilde{t})\frac{d\mu_j}{d\kappa} + f_x(X_{L0}(\tilde{t}) + \mu_0, \tilde{t})\frac{d\mu_0}{d\kappa}\mu_j = \lambda_{j-1}(\kappa),$$

where the λ_{j-1}'s are uniquely determined in terms of preceding coefficients, and they must have appropriate stability behavior as $\kappa \to \pm\infty$. In particular, since $\lambda_{j-1}(\kappa)$ decays exponentially as $\kappa \to -\infty$ and $\mu_j(-\infty) = 0$, an integration implies that

$$\frac{d\mu_j}{d\kappa} + f(X_{L0}(\tilde{t}) + \mu_0, \tilde{t})\mu_j = \int\limits_{-\infty}^{\kappa} \lambda_{j-1}(r)dr.$$

Since the corresponding homogeneous equation has the solution $d\mu_0/d\kappa = -\int_0^{\mu_0(\kappa)} f(X_{L0}(\tilde{t}) + r, \tilde{t})dr$, variation of parameters implies that μ_j must be of the form

$$\mu_j(\kappa) = \left(-\int\limits_0^{\mu_0(\kappa)} f(X_{L0}(\tilde{t}) + s, \tilde{t})ds \right) \left(\beta_j \right.$$

$$\left. + \int\limits_{\kappa}^{\infty} \frac{\left[\int_{-\infty}^p \lambda_{j-1}(r)dr \right] dp}{\int_0^{\mu_0(p)} f(X_{L0}(\tilde{t}) + s, \tilde{t})ds} \right).$$

Since the first factor tends to zero as $\kappa \to \infty$, the second factor must become infinite in order for $\mu_j(\kappa)$ to attain its prescribed limiting behavior. This limit should uniquely determine β_j.

Readers might wonder if we could analogously construct asymptotic solutions which switch more than once between limiting solutions which are (appropriately stable) solutions of the reduced equation. In the resulting interior intervals, there then would be boundary layers at both endpoints. This suggests that the function $f(X_0, t)$ will need to have isolated zeros (and thereby the full equation will have turning points) within the interval or that $f(X_0, t)$ be identically zero locally. In either case, the construction of any such limiting solutions would necessarily be different than the preceding.

Example

Lorenz (1984) considered the example

$$\begin{cases} \epsilon\ddot{x} + x(1 - x^2)\dot{x} - x = 0, \\ x(0) = 1.6, \qquad x(1) = -1.7 \end{cases}$$

which he showed to have a unique, numerically attainable, solution. Our preceding discussion makes it natural to seek a solution of the form

$$\lim_{\epsilon \to 0} x(t) = X_0(t) = \begin{cases} X_{L0}(t), & 0 \le t < t_1, \\ X_{S0}(t) \equiv 0, & t_1 < t < t_2, \\ X_{R0}(t), & t_2 < t \le 1, \end{cases}$$

where X_{L0} and X_{R0} satisfy the reduced problems

$$X_{L0}(1 - X_{L0}^2)\dot{X}_{L0} = X_{L0}, \qquad X_{L0}(0) = 1.6$$

and

$$X_{R0}(1 - X_{R0}^2)\dot{X}_{R0} = X_{R0}, \qquad X_{R0}(1) = -1.7,$$

respectively, while the singular limiting solution $X_{S0}(t) \equiv 0$ trivially satisfies the reduced equation. Cancelling the nonzero factor and integrating the resulting equations, we obtain the implicit outer solutions

$$X_{L0} - \tfrac{1}{3}X_{L0}^3 = t + C_{L0}$$

and

$$X_{R0} - \tfrac{1}{3}X_{R0}^3 = t + C_{R0}$$

on the intervals $0 \leq t < t_1$ and $t_2 < t \leq 1$, respectively. The constants then become specified by the respective prescribed endvalues. Any transition layer solution $\varphi(\kappa, \epsilon)$ near t_1 might naturally be described through the stretched variable

$$\kappa = (t - t_1)/\epsilon$$

and the layer equation

$$\frac{d^2\varphi}{d\kappa^2} + \varphi(1 - \varphi^2)\frac{d\varphi}{d\kappa} - \epsilon\varphi = 0.$$

(Instead of seeking the transition layer solution φ, we could instead correct the outer solution X_L locally.) The layer solution must match $X_L(t_1 + \epsilon\kappa, \epsilon)$ as $\kappa \to -\infty$ and $X_S(t_1 + \epsilon\kappa, \epsilon) \equiv 0$ as $\kappa \to \infty$. Integration from $+\infty$ implies that the leading term φ_0 should satisfy

$$\frac{d\varphi_0}{d\kappa} + F(\varphi_0) = 0,$$

where $F(\varphi) = \tfrac{1}{4}\varphi^2(2 - \varphi^2)$. Since $F(\varphi_0(\pm\infty)) = 0$, we will need $F = 0$ at the rest point $X_{L0}(t_1)$. Thus, $X_{L0}(t_1) = 0$ or $\pm\sqrt{2}$, and stability of X_{L0} requires $f(X_{L0}) = X_{L0}(1 - X_{L0}^2)$ to be negative on $0 \leq t < t_1$. This determines $X_{L0}(t_1) = \sqrt{2}$, so

$$t_1 = \frac{\sqrt{2}}{3} - 1.6 + \frac{1}{3}(1.6)^3 \approx 0.237.$$

Analogously, on $t_2 < t < 1$, $X_{R0}(1 - X_{R0})^2$ must be positive. Integrating the layer equation at t_2 and matching to the singular solution now requires us to have $X_{R0}(t_2) = -\sqrt{2}$, so

$$t_2 = -\frac{\sqrt{2}}{3} - \frac{1}{3}(1.7)^3 + 2.7 \approx 0.591.$$

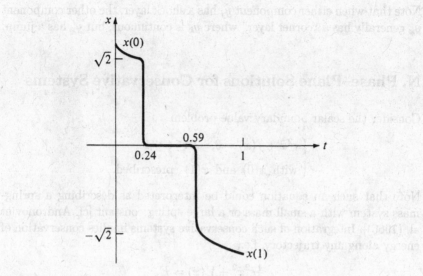

Fig. 3.8. The solution to $\epsilon\ddot{x} + x(1 - x^2)\dot{x} - x = 0$, $x(0) = 1.6$, $x(1) = -1.7$.

Further analysis would be necessary to prove that this formally constructed limiting solution is asymptotically correct. In particular, we need to show that the trivial limiting solution within (t_1, t_2) is appropriately stable. Howes (1978) considers natural stability hypotheses for such singular limits. Lorenz (1984) guaranteed the uniqueness and correctness of the result, based on the numerical convergence of an asymptotically correct approximation obtained by using an "upwinded" finite-difference scheme. The solution obtained is pictured in Figure 3.8. We note that this example has a rich variety of asymptotic solutions as we let the endvalues $x(0)$ and $x(1)$ vary [cf. Lorenz (1982a, b), Allen and O'Malley (1990), and O'Malley (1991)].

Exercise

[Compare O'Malley (1983) and Kelley (1988).]

Consider the vector two-point problem

$$\begin{cases} \epsilon y_1'' + y_1 y_1' - \tfrac{1}{5} y_1(-3y_1 + y_2) = 0, & y_1(0) = -5, y_1(1) = 3, \\ \epsilon y_2'' + y_2 y_2' - \tfrac{1}{5} y_2(-20y_1 + 6y_2) = 0, & y_2(0) = -5, y_2(1) = 12. \end{cases}$$

a.   Construct an asymptotic solution for which y_1 has an initial layer and y_2 a terminal layer.

b.   Construct another asymptotic solution for which y_1 has a shock layer at some x_1 and y_2 a shock layer at some x_2 with $0 < x_1 < x_2 < 1$.

c.   Construct a third asymptotic solution for which y_1 has a shock layer at some x_3 and y_2 one at some x_4 for $0 < x_4 < x_3 < 1$.

Note that when either component y_j has a shock layer, the other component y_k generally has a "corner layer" where y_k is continuous, but y_k' has a jump.

N. Phase–Plane Solutions for Conservative Systems

Consider the scalar boundary value problem

$$\begin{cases} \epsilon^2 \ddot{x} + f(x) = 0, & 0 \le t \le 1 \\ \text{with } x(0) \text{ and } x(1) \text{ prescribed.} \end{cases}$$

Note that such an equation could be interpreted as describing a spring-mass system with a small mass or a large spring constant [cf. Andronov et al. (1966)]. Integration of such conservative systems implies conservation of energy along any trajectory, i.e.,

$$\tfrac{1}{2}\epsilon^2 \dot{x}^2 + V(x) = E,$$

where E is the constant total energy and $V(x) = \int^x f(s)ds$ is the potential energy. Introducing

$$z = \epsilon \dot{x}$$

[noting that x and z are Diener's (1981) "plane of observability" variables], it becomes natural to consider such systems in the x–z phase plane [cf. Minorsky (1947)] where we have the singularly perturbed system

$$\begin{cases} \epsilon \dot{x} = z, \\ \epsilon \dot{z} = -f(x). \end{cases}$$

The relationship

$$z = \pm\sqrt{2[E - V(x)]}$$

allows us to draw any possible trajectory. We first graph $V(x)$. Then, for the appropriate total energy E, we can draw the symmetric, real $z(x)$ profiles as long as $V(x)$ does not exceed E. See Figures 3.9 and 3.10. Moreover, given any two end values $x(0)$ and $x(1)$, we can attempt to adjust the energy level E in order to determine a (bounded) trajectory $z(x)$ along which

$$\epsilon \int\limits_{x(0)}^{x(1)} \frac{dx}{z(x)} = 1.$$

This is naturally imposed since $dt = \epsilon(dx/z)$. As $\epsilon \to 0$, then, z cannot remain bounded away from zero on any such path. Thus, E must tend asymptotically to a maximum value of V. We note that the corresponding rest points in the x–z plane are saddle points when $f' < 0$. (If f' were zero, we would have a higher-order rest point.) For most t in $(0,1)$, we can

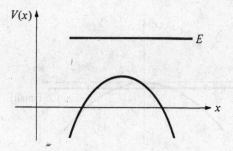

Fig. 3.9. Potential energy $V(x)$.

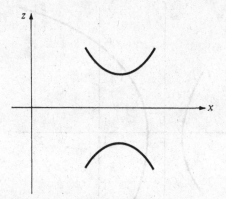

Fig. 3.10. Graphical determination of the corresponding $z(x)$.

expect the solution x of our two-point problem to remain near roots of f corresponding to such maxima. Other rest points (in particular, minima of V) are largely irrelevant. A study of the complete phase portrait provides all possible solutions [cf. O'Malley (1990) for more details].

Example 1

$$\begin{cases} \epsilon^2 \ddot{x} - x = 0, \\ x(0) = 1, x(1) = 2. \end{cases}$$

Here, $V(x) = -\frac{1}{2}x^2$ has a unique maximum at $x = 0$, so the solution is asymptotically zero, except in the necessary endpoint layers. See Figures 3.11 and 3.12. Consider a small negative value of E. The corresponding right arc of $z(x)$ provides a trajectory $\alpha\beta$ along which the solution can move rapidly from $x = 1$ to x nearly zero, where it lingers, and it then

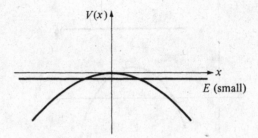

Fig. 3.11. The potential $V(x) = -\frac{1}{2}x^2$.

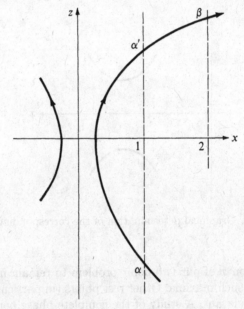

Fig. 3.12. Graphical solution of $\epsilon^2\ddot{x} = x$, $x(0) = 1$, $x(1) = 2$.

moves rapidly from $x = 0$ to $x = 2$. (The shorter monotonic trajectory $\alpha'\beta$ is not satisfactory because it is transversed too rapidly.) Thus, we will have Figure 3.13. We note that the same approach does not provide a solution of the two-point problem for $\epsilon^2\ddot{x} + x = 0$, because the potential $V(x) = -\frac{1}{2}x^2$ has no finite maxima.

Example 2

The problem

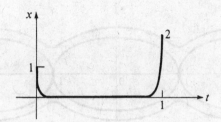

Fig. 3.13. The solution $x(t, \epsilon)$ for small ϵ.

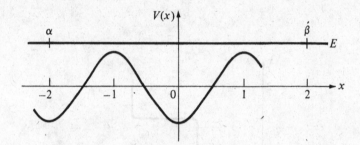

Fig. 3.14. The potential $V(x) = -(1/\pi)\cos(\pi x)$.

$$\begin{cases} \epsilon^2 \ddot{x} + \sin(\pi x) = 0, \\ x(0) = -2, \qquad x(1) = 2 \end{cases}$$

has the potential energy $V(x) = -(1/\pi)\cos(\pi x)$. We can solve the boundary value problem using an energy E slightly greater than $1/\pi$. Thus, we will have Figures 3.14 and 3.15. The desired trajectory will lie close to, but above, the separatrix which passes through the saddle points $(\pm 1, 0)$. Since motion is fast away from the rest points, most time will be spent near $x = -1$ or $x = 1$. Because z behaves the same near $x = \pm 1$, the time available will be divided equally between the two limiting values. The resulting monotonically increasing solution will be unique because of this phase plane analysis (see Figure 3.16).

If we, instead, had the endvalues $x(0) = -2$ and $x(1) = 1$, two-thirds of the time will be spent near $x = -1$ and only one-third of the time would be spent near $x = 1$. Thus, we would have Figure 3.17.

We note that detailed construction of the layer solution near the boundary and interior layers could be carried out. The location of the layers is not arbitrary, but completely specified by the end values.

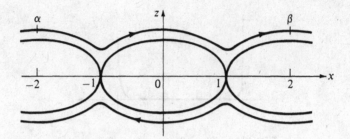

Fig. 3.15. Graphical solution of $\epsilon^2 \ddot{x} = -\sin(\pi x)$, $x(0) = -2$, $x(1) = 2$.

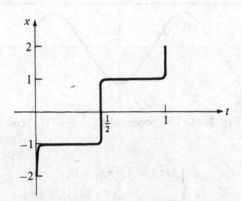

Fig. 3.16. The solution for ϵ small.

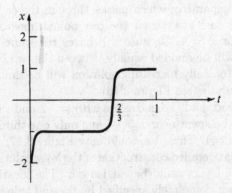

Fig. 3.17. The solution with $x(0) = -2$ and $x(1) = 1$.

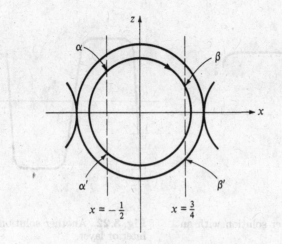

Fig. 3.18. Possible solution trajectories.

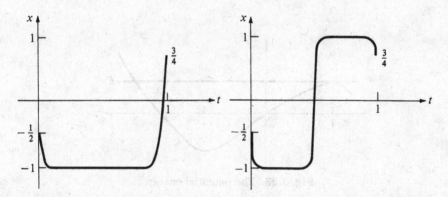

Fig. 3.19. A solution with two end-point layers.

Fig. 3.20. A solution with an interior layer.

Example 3

$$\begin{cases} \epsilon^2 \ddot{x} + \sin(\pi x) = 0, \\ x(0) = -\tfrac{1}{2}, \qquad x(1) = \tfrac{3}{4}. \end{cases}$$

To allow the trajectory to approach a rest point (either ± 1), we must use an energy level slightly less than $1/\pi$. The trajectory will lie within the separatrix encircling the origin. See Figure 3.18. Let α and α' denote the crossings of the orbit and the line $x(0) = -\tfrac{1}{2}$ and let β and β' denote its crossings with $x(1) = \tfrac{3}{4}$. Note that the trajectory $\alpha\beta$ cannot be used since its passage only takes $O(\epsilon)$ time. Possible trajectories include

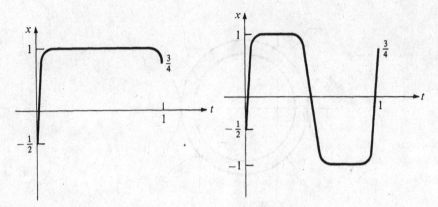

Fig. 3.21. Another solution with an endpoint layer.

Fig. 3.22. Another solution with an interior layer.

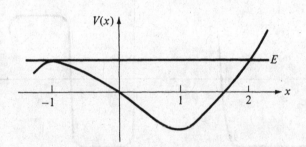

Fig. 3.23. The potential energy.

$$\alpha'\alpha\beta, \qquad \alpha'\alpha\beta\beta', \ldots$$

and

$$\alpha\beta\beta', \qquad \alpha\beta\beta'\alpha'\alpha\beta, \ldots.$$

The corresponding solution graphs are represented in Figures 3.19 through 3.22. Thus, solutions can have any number of jumps, depending on how many times the rest points are approached. Further, their spacing of jumps is regular. Note that this is reminiscent, to some extent, of "chattering," which occurs in time-optimal control theory [cf. Utkin (1978)]. This approach also explains why certain formally constructed "solutions" are spurious [cf. Carrier and Pearson (1968)].

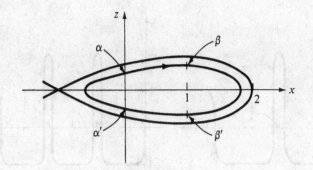

Fig. 3.24. Graphical construction of solutions to $\epsilon^2 \ddot{x} + x^2 - 1 = 0$, $x(0) = 0$, $x(1) = 1$.

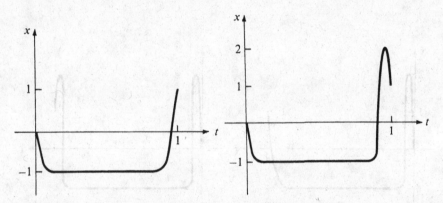

Fig. 3.25. A solution with endpoint layers.

Fig. 3.26. Another solution with endpoint layers.

Example 4

For the problem

$$\begin{cases} \epsilon^2 \ddot{x} + x^2 - 1 = 0, \\ x(0) = 0, \qquad x(1) = 1, \end{cases}$$

the potential energy $V(x) = \frac{1}{3}x^3 - x$ has a maximum at $x = -1$. We note that $V(2) = V(-1)$, but $x = 2$ is not an extreme point. We have Figures 3.23 and 3.24. The phase plane has a separatrix for $E = V(-1)$. To connect $x(0) = 0$ and $x(1) = 1$ with a trajectory passing the rest point, we need E slightly less than $V(-1)$.

There will be an infinite number of solutions to our boundary value problem. If we let α and α' denote the locations where the trajectory crosses $x = 0$ and β and β' the crossings of $x = 1$, possible trajectories include

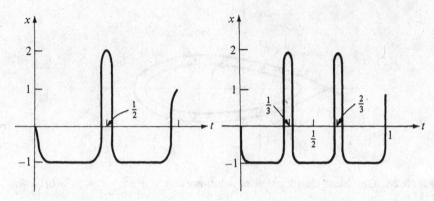

Fig. 3.27. A one spike solution.   **Fig. 3.28.** A two spike solution.

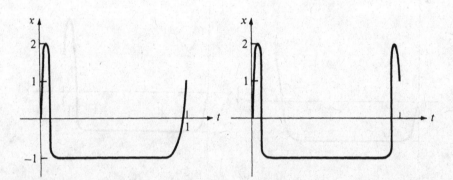

Fig. 3.29. An endpoint spike solution.   **Fig. 3.30.** A solution with two endpoint spikes.

$$\alpha'\alpha\beta, \quad \alpha'\alpha\beta\beta', \quad \alpha'\alpha\beta\beta'\alpha'\alpha\beta, \quad \alpha'\alpha\beta\beta'\alpha'\alpha\beta\beta'\alpha'\alpha\beta, \ldots$$

and

$$\alpha\beta\beta'\alpha'\alpha\beta, \quad \alpha\beta\beta'\alpha'\alpha\beta\beta', \ldots.$$

The corresponding solution graphs are given respectively by Figures 3.25 through 3.30. The solutions, then, all tend to $X_0(t) = -1$ for almost all t. There are, however, endpoint layers (which are not all monotonic) and an arbitrary number of regularly-spaced interior "spikes" where the solution jumps rapidly from $x = -1$ to 2.

We note that related phenomena for some nonautonomous problems have been reported by Lutz and Goze (1981), Lange (1983), and Kath (1985). Computations presented in Bender and Orszag (1978) for the equation

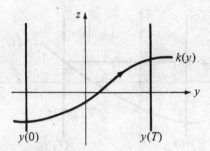

Fig. 3.31. Liénard plane trajectory.

$$\epsilon^2 y'' = 1 - 2(1 - x^2)y - y^2$$

show that an infinite number of solutions with regularly spaced spikes (and nonmonotonic boundary layers) can occur for nonautonomous equations. (Their pictures are, incidentally, somewhat misleading concerning the asymptotic limit because they are carried out for the relatively large value $\epsilon = \sqrt{0.1}$.) The analytical results of Kath et al. (1987) also illustrate related conclusions by another method [cf. Ward (1991) as well].

O. A Geometric Analysis for Some Autonomous Equations

Now consider the autonomous scalar problem

$$\begin{cases} \epsilon \ddot{y} + [F(y)] \dot{}\, + h(y) = 0 \quad \text{on } 0 \le t \le T \\ \text{where } y(0) = A \text{ and } y(T) = B \text{ are prescribed.} \end{cases}$$

Here, F and h are assumed to be smooth and A, B, and T are allowed to vary. Such problems were discussed in Lutz and Goze (1981) and by others interested in applying nonstandard analysis to singular perturbation problems [cf. Diener and Diener (1987)]. Our discussion will be intuitive, aiming to simply obtain limiting behavior. To proceed, let us introduce the Liénard variable

$$z = \epsilon \dot{y} + F(y),$$

thereby obtaining the fast–slow system

$$\begin{cases} \epsilon \dot{y} = z - F(y), \quad y(0) \text{ and } y(T) \text{ prescribed,} \\ \dot{z} = -h(y). \end{cases}$$

We note that Liénard (1928) used such transformations to study nonlinear oscillations described by initial value problems and that the slow variable

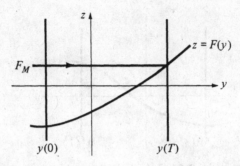

Fig. 3.32. Example where F has an endpoint maximum.

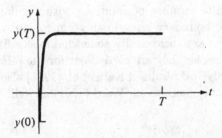

Fig. 3.33. Solution corresponding to the trajectory in Figure 3.32.

$z = -\int^t h(y(s))ds$ will be smoother than the fast variable y (see Figure 3.31). In the y–z plane, then, we are seeking a trajectory which passes from some point on the vertical line $y = A$ to a point on the vertical line $y = B$, taking T units of "time" t. If the trajectory can be represented by a smooth function $z = k(y), dt = -dz/h$ implies that

$$T = \int_B^A \frac{k'(y)}{h(y)} dy.$$

Note that motion will be rapid [since $\dot{y} = O(1/\epsilon)$ will be large] and nearly horizontal [since $dz/dy = O(\epsilon)$], except within an $O(\epsilon)$ neighborhood of the *characteristic curve* $\Gamma : z = F(y)$. Moreover, y will increase above Γ and decrease below Γ, whereas z will increase (or decrease) for $h < 0$ (or $h > 0$). In order to use up the allotted time, bounded trajectories must usually follow arcs of Γ, sometimes remaining near rest points for quite some while. Along $\Gamma, Z = F(Y)$ implies that Y will satisfy the reduced problem

$$f(Y)\dot{Y} + h(Y) = 0$$

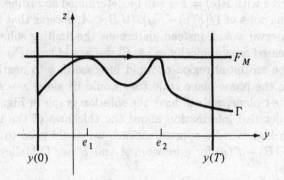

Fig. 3.34. Example when F has two equal maxima.

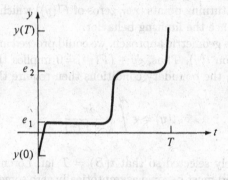

Fig. 3.35. Solution corresponding to the trajectory of Figure 3.34.

where $f(Y) = F'(Y)$. Near points $\tilde{t}$ of rapid change, a new variable $(t - \tilde{t})/\epsilon$ will be appropriate for describing the fast motion, provided appropriate hypotheses on F hold in the layer. Layers with other thicknesses may, however, sometimes be required. Note that the rest points in the y–z system occur on Γ wherever $h(y) = 0$. Moreover, the maximum principle guarantees that a unique solution will exist whenever $h'(y) < 0$ everywhere.

Example 1 $h(y) \equiv 0$.

In this situation, z will be constant and all points of Γ will be rest points. Trajectories must pass above Γ if $B > A$, and below Γ if $B < A$.

For definiteness, suppose $B > A$. Then, z must remain slightly above (but asymptotically equal to) the maximum F_M of F in $[A, B]$. (A smaller value of z would not allow the trajectory to reach B in a finite time T, whereas a larger value of z would take less time.) Since $\dot{y} > 0$, y will increase monotonically. As $\epsilon \to 0$, the proportionate amount of time spent near each point e with $F(e) = F_M$ will be determined according to the local strengths of the poles of $1/[F(y) - F_M]$. (If $B < A$, observe that the minima of F in the interval would, instead, determine the limiting solution.)

If F increased monotonically in $[A, B]$, we would have $F_M = F(B)$, so there would be an initial region of rapid increase in y to nearly the level B. Pictorially, the phase-plane trajectory would lie along $z \sim F_M$ (Figure 3.32). In the y–t plane, we will have the solution graph in Figure 3.33. To obtain more detailed information about the thickness of the initial layer and to obtain higher-order approximations, we would need to specify the behavior of $F(B) - F(y)$ for y between A and B [see O'Malley (1991) for certain details].

If, instead, there were a finite number of values e_i in the interval (A, B) with $F(e_i) = F_M$, the limiting solution $Y_0(t)$ would be a monotonically increasing step function with successive values e_i. Pictorially, for two such maxima, we would have Figures 3.34 and 3.35.

We note that turning points (i.e., zeros of $F'(y)$) which are not maxima of F do not influence the limiting behavior.

Instead of this geometric approach, we could proceed analytically, using the inverse function $t(y)$. Thus, $\epsilon \ddot{y} + (F(y)) \dot{=} 0$ implies that $\epsilon \dot{y} + F(y)$ is constant. Further, the boundary conditions then require that

$$t(y) = \epsilon \int\limits_{A}^{y} \frac{dr}{k - F(r)},$$

where k is uniquely selected so that $t(B) = T$ [cf. Lorenz (1982b)]. This shows that $k - F(y)$ must become asymptotically zero somewhere along any finite trajectory.

Example 2 $(F(y) = \frac{1}{2}y^2, \qquad h(y) = -y)$

Here we have the Lagerstrom–Cole model problem: $\epsilon \ddot{y} + y\dot{y} - y = 0$. It has a unique solution for every $\epsilon > 0$. The corresponding fast–slow system is given by

$$\begin{cases} \epsilon \dot{y} = z - \frac{1}{2}y^2, \\ \dot{z} = y. \end{cases}$$

Representative trajectories in the y–z phase plane are pictured in Figure 3.36.

They could readily be obtained on a graphics terminal by integrating forward and backward in time from well-chosen initial points. Note that the portion Γ^- of Γ with $y < 0$ is repellent, in the sense that trajectories ultimately leave it, whereas that for $y > 0, \Gamma^+$, is attractive.

As long as a trajectory continues to follow Γ^-, we will have

$$Y\dot{Y} = Y, \qquad Y(0) = A < 0.$$

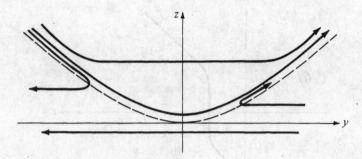

Fig. 3.36. Liénard plane trajectories for the Lagerstrom–Cole model equation.

So, if Y remains nonzero, we will have

$$Y = Y_L(t) = t + A.$$

Likewise, for a trajectory which ends along Γ^+, we will have $Y\dot{Y} = Y$ and $Y(T) = B > 0$, so

$$Y = Y_R(t) = t + B - T.$$

On Γ, when $\dot{Y} = 1$ and $Y \neq 0$, we will have $dt = dy$.

   If we then seek a solution which jumps from Y_L to Y_R at some $\tilde{t}$ in $(0,1)$, the rapid transition will be represented by a horizontal motion in this phase plane, moving from Γ^- to Γ^+. Thus, the symmetry of Γ requires

$$Y_L(\tilde{t}) = -Y_R(\tilde{t}) < 0.$$

This coincides with the classical Rankine–Hugoniot condition. Moreover, it determines the jump location $\tilde{t}$ since $\tilde{t} + A = -(\tilde{t} + B - T)$ implies that

$$\tilde{t} = \tfrac{1}{2}(T - A - B).$$

Pictorially, we will have Figure 3.37.

   To guarantee that $0 < \tilde{t} < T$, we must restrict A and B to have

$$-T < A + B < T.$$

To guarantee that $Y_L(\tilde{t}) < 0$ further requires that $\tilde{t} + A = \tfrac{1}{2}(T + A - B) < 0$, so we also need

$$T < B - A.$$

This shows that such a shock layer can only occur if A and B are restricted to the shaded region pictured in the $A - B$ plane of Figure 3.38. For all other A, B, and T values, the limiting solutions can also be found [cf. Kevorkian and Cole (1981)]. For example, in the triangle where $A < 0 < B$ and $T > B - A$, the phase portrait suggests that there is a trajectory which follows Γ_- to a small neighborhood of the origin stays there awhile, and then ultimately follows Γ^+. The time required as y varies from A to 0

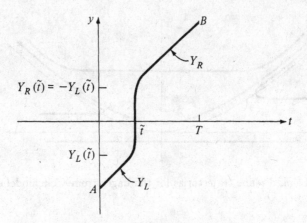

Fig. 3.37. A shock-layer solution.

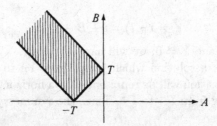

Fig. 3.38. The shock-layer region in the $A - B$ plane.

along Γ^- is $-A$ and that needed to traverse Γ^+ from $y = 0$ to B is B. The remaining time, $T - B + A$, is spent asymptotically near the rest point, so the limiting solution is given by Figure 3.39, i.e., the trivial singular solution of the reduced problem provides the limiting solution for $-A < t < T - B$. Observe, again, that this limit has been found without doing any stretching and matching. To determine a more complete asymptotic solution, such details are, of course, unavoidable.

Example 3   [Lorenz (1982)]

The equation

$$\epsilon \ddot{y} + y(1 - y^2)\dot{y} - y = 0$$

corresponds to the system

$$\begin{cases} \epsilon \dot{y} = z - F(y), \\ \dot{z} = y \end{cases}$$

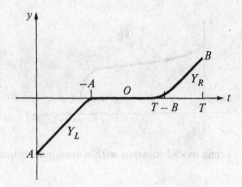

Fig. 3.39. Solution with two corner layers.

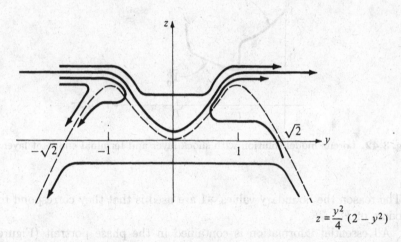

$$z = \frac{y^2}{4}(2 - y^2)$$

Fig. 3.40. Liénard plane trajectories for the Lorenz model equation.

with $F(y) = \frac{1}{4}y^2(2-y^2)$. It can be shown (using the theory of Section 3M) that the two-point problem has a unique solution for all values of A, B, and T, for any $\epsilon > 0$. We note that the reduced equation

$$Y_0(1 - Y_0^2)\dot{Y}_0 - Y_0 = 0$$

has the singular solution $Y_0(t) \equiv 0$ as well as solutions of $(1 - Y_0^2)\dot{Y}_0 = 1$, i.e. $Y_0 - \frac{1}{3}Y_0^3 = t + C$ for some constant C. A detailed study by John Allen shows that for all A, B, and T, the limiting solution as $\epsilon \to 0$ jumps between the trivial singular solution and the reduced solutions satisfying the implicit relation $Y_0 - \frac{1}{3}Y_0^3 = t + C_i$, for $i = 1, 2, 3$, and 4, determined by one of the four boundary conditions $Y_0(0) = A$ or -1 or $Y_0(T) = B$ or

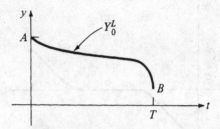

Fig. 3.41. Lorenz model solution with a terminal endpoint layer.

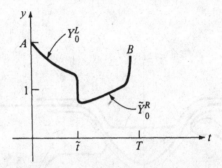

Fig. 3.42. Lorenz model solution with shock layer and terminal endpoint layer.

1. The reason the boundary values ∓ 1 are used is that they correspond to maxima of F.

All essential information is contained in the phase portrait (Figure 3.40). Consider, for example, the case that $A > 1$ and $B > 1$. The limiting solution Y_0^L obtained is defined by $(1 - Y_0^2)\dot{Y}_0 = 1, Y_0(0) = A$. It holds except in a terminal boundary layer. Pictorially, this yields Figure 3.41. The time used is determined by integrating $dt = dz/y = (1 - y^2)dy$ along the trajectory. Thus, the maximum time of passage along such a trajectory is $T_1 \equiv \int_A^1 (1 - y^2)dy$.

For $T > T_1$, one would instead again travel along Y_0^L, then jump horizontally left in the Liénard plane from Y_0^L to $\tilde{Y}_0^R$ [defined by $(1 - Y_0^2)\dot{Y}_0 = 1, Y_0(T) = 1$] at some value $\tilde{t}$ where $1 < Y_0^L(\tilde{t}) < \sqrt{2}$, follow $\tilde{Y}_0^R$ until y becomes 1 at nearly time T, and, finally, jump rapidly and horizontally right from $\tilde{Y}_0(T) = 1$ to B. In the y–t plane, we would have Figure 3.42.

Note that the jump from Y_0^L to $\tilde{Y}_0^R$ at $\tilde{t}$ will satisfy the Rankine–Hugoniot condition $F(Y_0^L(\tilde{t})) = F(\tilde{Y}_0^R(\tilde{t}))$. Moreover, the time needed along the full trajectory is given by

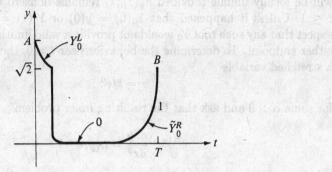

Fig. 3.43. Lorenz model solution with shock layer, corner layer, and terminal endpoint layer.

$$\int_A^{Y_0^L(t)} (1-y^2)dy + \int_{Y_0^R(t)}^1 (1-y^2)dy.$$

When $A > \sqrt{2}$ and $T > T_2 \equiv \int_A^{\sqrt{2}}(1-y^2)dy + \int_0^1(1-y^2)dy$, we must instead follow $Y_0^L(t)$ from $y = A$ to $y = \sqrt{2}$, then jump to near the rest point at the origin, remain there for some time, and, finally, follow $\tilde{Y}_0^R(t)$ from $y = 0$ to $y = 1$, before jumping horizontally in the Liénard plane to B. Pictorially, in the y–t plane, we will have Figure 3.43.

Exercise

[Compare Lorenz (1982).]

Determine the limiting solution to the boundary value problem

$$\epsilon\ddot{y} + y(1-y^2)\dot{y} - y = 0, \quad y(0) = \tfrac{3}{2}, \quad y(1) = \delta$$

for all values of δ.

P. Semilinear Problems

We will now consider scalar problems of the form

$$\epsilon^2 y'' = h(y,t), \qquad 0 \le t \le 1,$$

with endvalues $y(0)$ and $y(1)$ prescribed, for a smooth function h. We note that any solution Y_0 of the reduced problem

$$h(Y_0, t) = 0$$

will be locally unique provided $h_y(Y_0, t)$ remains nonzero throughout $0 \leq t \leq 1$. Unless it happened that $Y_0(0) = y(0)$ or $Y_0(1) = y(1)$, we must expect that any such root Y_0 would not provide a valid limiting solution near either endpoint. To determine the behavior near $t = 0$, then, we introduce a stretched variable

$$\tau = t/\epsilon^\alpha$$

for some $\alpha > 0$ and ask that the resulting inner problem

$$\epsilon^{2(1-\alpha)} \frac{d^2 y}{d\tau^2} = h(y, \epsilon^\alpha \tau)$$

have a limiting solution $\tilde{y}(\tau)$ on the semi-infinite interval $\tau \geq 0$ which has the initial value $\tilde{y}(0) = y(0)$ and the steady-state limit $Y_0(0)$ as $\tau \to \infty$. Since h will generally be nontrivial in this initial layer, we must take $\alpha = 1$, thereby retaining some limiting dynamics in the layer. Note that the $O(\epsilon)$ initial layer is thicker than obtained for our quasilinear problems (considering that the coefficient of y'' is now ϵ^2, rather than ϵ). The boundary value problem for $\tilde{y}(\tau)$ can be integrated once (from infinity) to yield

$$\frac{1}{2} \left(\frac{d\tilde{y}}{d\tau} \right)^2 = \int_{Y_0(0)}^{\tilde{y}(\tau)} h(r, 0) dr.$$

Note that this can define a real solution only if

$$\int_{Y_0(0)}^{y} h(r, 0) dr > 0$$

holds for all y between $y(0)$ and $Y_0(0)$. We will refer to this assumption as the *initial layer stability condition*. Since we already assumed that $h_y(Y_0, t) \neq 0$, a Taylor series approximation of the integral about $y = Y_0(0)$ implies that we must have $h_y(Y_0(0), 0) > 0$. Thus,

$$h_y(Y_0(t), t) > 0$$

throughout $0 \leq t \leq 1$. We will call this condition *stability of the limiting solution $Y_0(t)$*. We note that the integral condition implies that we must have

$$\frac{d\tilde{y}}{d\tau} = -\text{sgn}[y(0) - Y_0(0)] \sqrt{2 \int_{Y_0(0)}^{\tilde{y}} h(r, 0) dr}, \quad \tilde{y}(0) = y(0),$$

on $\tau \geq 0$ [where $\text{sgn}(z)$ is the sign of z]. Although we cannot generally find $\tilde{y}$ explicitly, our integral condition guarantees that the solution will be unique, will exist for all $\tau \geq 0$, and that $|\tilde{y}(\tau) - Y_0(0)|$ will decay monotonically to zero as $\tau \to \infty$. Performing an analogous asymptotic argument near

$t = 1$ shows that we could naturally impose the corresponding *terminal layer stability condition* that

$$\int_{Y_0(1)}^{y} h(r,1)dr > 0$$

for all y between $y(1)$ and $Y_0(1)$.

Example 1

The exact solutions of the linear equation

$$\epsilon^2 y'' = y$$

are given by $y(t) = e^{-t/\epsilon}C_1 + e^{-(1-t)/\epsilon}C_2$ for arbitrary constants C_1 and C_2. Thus, Dirichlet boundary conditions would determine the unique solution

$$y(t) = e^{-t/\epsilon}\left(\frac{y(0) - y(1)e^{-1/\epsilon}}{1 - e^{-2/\epsilon}}\right) + e^{-(1-t)/\epsilon}\left(\frac{y(1) - y(0)e^{-1/\epsilon}}{1 - e^{-2/\epsilon}}\right).$$

Hence, the limiting solution as $\epsilon \to 0$ is trivial within (0,1) while the $O(\epsilon)$ thick endpoint layer corrections agree with the uniform approximation

$$y(t) \sim e^{-t/\epsilon}y(0) + e^{-(1-t)/\epsilon}y(1)$$

which is valid to all orders $O(\epsilon^N)$.

Example 2

The Dirichlet problem for the equation

$$\epsilon^2 y'' = -y$$

has the exact solution

$$y(t) = \left(\cos\frac{t}{\epsilon}\right)y(0) + \frac{\sin(t/\epsilon)}{\sin(1/\epsilon)}\left[y(1) - \left(\cos\frac{1}{\epsilon}\right)y(0)\right]$$

provided $\sin(1/\epsilon) \neq 0$. Since the solution is undefined for a sequence $\{\epsilon_n\}$ of decreasing ϵ values tending to zero, the problem has no limiting solution as $\epsilon \to 0$. Seeking an initial layer correction is actually inappropriate for such rapidly oscillating solutions. Indeed, our usual initial layer stability condition that $\int_0^y h(r,0)dr = -\int_0^y rdr = -y^2/2 > 0$ for $y \neq 0$ cannot be satisfied.

Example 3

For the nonlinear equation

$$\epsilon^2 y'' = y^2,$$

the reduced equation has only the trivial solution $Y_0(t) = 0$. Decaying solutions of the corresponding limiting initial layer problem

$$\frac{d^2 \tilde{y}}{d\tau^2} = \tilde{y}^2$$

must satisfy

$$\frac{1}{2}\left(\frac{d\tilde{y}}{d\tau}\right)^2 = \frac{1}{3}\tilde{y}^3$$

for $\tilde{y}$ between $y(0)$ and 0. To get such a solution, we must ask that $y(0)$ be non-negative. {We could also show nonexistence by a convexity argument [cf. Chang and Howes (1984)]}. Then, however, we can uniquely integrate the initial value problem for $d\tilde{y}/d\tau = -\sqrt{\frac{2}{3}}\tilde{y}^{3/2}$ to see that the limiting layer is described by the algebraically decaying function

$$\tilde{y}(\tau) = \frac{y(0)}{1 + \sqrt{[2y(0)/3]\tau + \frac{1}{6}y(0)\tau^2}}.$$

Likewise, we can similarly obtain a terminal layer correction provided $y(1) \geq 0$.

Returning to the general equation

$$\epsilon^2 y'' = h(y, t),$$

let us select a solution $Y_0(t)$ of the reduced problem which satisfies

$$h_y(Y_0(t), t) > 0$$

throughout $0 \leq t \leq 1$. Our preceding discussion suggests that we seek a uniform representation of the corresponding asymptotic solution in the form

$$y(t) = Y(t, \epsilon) + \xi(\tau, \epsilon) + \eta(\sigma, \epsilon),$$

where the outer solution Y, the initial layer correction ξ, and the terminal layer correction η all have power series expansions in ϵ. We will also suppose that the correction $\xi \to 0$ as $\tau = t/\epsilon \to \infty$, that $\eta \to 0$ as $\sigma = (1-t)/\epsilon \to \infty$, and that $Y(t, 0) = Y_0(t)$.

Necessarily, the outer solution Y will be a smooth solution of the full system

$$\epsilon^2 Y'' = h(Y, t).$$

Setting

$$Y(t,\epsilon) \sim \sum_{j=0}^{\infty} Y_j(t)\epsilon^j,$$

we will have equality when $\epsilon = 0$ since we selected Y_0 as a root of $h(Y_0, t) = 0$. Coefficients of later powers ϵ^j imply that

$$h_y(Y_0(t), t)Y_j(t) = \beta_{j-1}(t),$$

where β_{j-1} is known successively. Since h_y is positive along Y_0, the Y_j's are uniquely determined termwise. Indeed, since the odd β_{j-1}'s are zero, the outer solution is actually a power series in ϵ^2.

Knowing $Y(t, \epsilon)$ asymptotically, the corresponding initial layer correction ξ must be a decaying solution of the nonlinear system

$$\frac{d^2\xi}{d\tau^2} = h(Y(\epsilon\tau, \epsilon) + \xi, \epsilon\tau) - h(Y(\epsilon\tau, \epsilon), \epsilon\tau)$$

on $\tau \geq 0$, which also satisfies the initial condition

$$\xi(0, \epsilon) = y(0) - Y(0, \epsilon).$$

Thus, its leading term ξ_0 must satisfy the nonlinear equation

$$\frac{d^2\xi_0}{d\tau^2} = h(Y_0(0) + \xi_0, 0) - h(Y_0(0), 0) = h(Y_0(0) + \xi_0, 0)$$

and later terms ξ_j must satisfy linear equations

$$\frac{d^2\xi_j}{d\tau^2} = h_y(Y_0(0) + \xi_0, 0)\xi_j + \gamma_{j-1}(\tau)$$

where γ_{j-1} is a successively known decaying function. Assuming that

$$\int_{Y_0(0)}^{y} h(r, 0)dr > 0$$

for all y between $y(0)$ and $Y_0(0)$, we can obtain ξ_0 uniquely as the solution of the initial value problem

$$\frac{d\xi_0}{d\tau} = -\text{sgn}[\xi_0(0)]\sqrt{2\int_0^{\xi_0} h(Y_0(0) + s, 0)ds}, \quad \xi_0(0) = y(0) - Y_0(0).$$

We could generally obtain ξ_0, without difficulty, through a numerical integration. Its ultimate exponential decay to zero follows from the positivity of $h_y(Y_0(0), 0)$. Because $d\xi_0/d\tau$ satisfies the homogeneous equation $d^2\eta/d\tau^2 = h_y(Y_0(0) + \xi_0, 0)\eta$, we can obtain a decaying solution of $d^2\zeta/d\tau^2 = h_y(Y_0(0) + \xi_0, 0)\zeta + q$ on $\tau \geq 0$ by variation of parameters. Thus,

$$\zeta(\tau) = \frac{d\xi_0}{d\tau} \left(\frac{\zeta(0)}{\frac{d\xi_0(0)}{d\tau}} - \int_0^\tau \frac{\left(\int_u^\infty \frac{d\xi_0}{d\tau}(r)q(r)dr \right) du}{\left(\frac{d\xi_0(u)}{d\tau} \right)^2} \right)$$

or

$$\zeta(\tau) = \sqrt{\int_0^{\xi_0(\tau)} h(Y_0(0) + s, 0)ds} \left(\frac{\zeta(0)}{\sqrt{\int_0^{\xi_0(0)} h(Y_0(0) + s, 0)ds}} \right.$$

$$\left. - \int_0^\tau \int_u^\infty \frac{\sqrt{\int_0^{\xi_0(r)} h(Y_0(0) + p, 0)dp}}{\int_0^{\xi_0(u)} h(Y_0(0) + p, 0)dp} q(r)dr \, du \right)$$

provided q decays sufficiently at infinity to guarantee the decay of ζ. In this way, we can successively obtain the boundary layer correction terms $\xi_j(\tau)$. The terminal layer correction at $t = 1$ can be obtained analogously provided

$$\int_{Y_0(1)}^y h(r, 1)dr > 0$$

for all y between $y(1)$ and $Y_0(1)$. That the formally obtained expansion is asymptotically valid can be shown by relatively straightforward methods [cf., e.g. Eckhaus (1979), Smith (1985), and Hale and Sakamoto (1988)].

If the reduced problem

$$h(Y_0, t) = 0$$

has more than one root which is stable throughout $0 \le t \le 1$, and the corresponding boundary layer problems also have appropriate stability, we might find more than one solution to the given Dirichlet problem. Maximum principle arguments can be used to show that the solution is unique, however, in any domain where the inequality $h_y > 0$ is maintained.

Example

Consider the two-point problem for the equation

$$\epsilon^2 \frac{d^2 y}{dt^2} = h(y, t) \equiv 1 + p(t)y - y^2$$

on an interval $0 \le t \le 1$ where $p(t)$ is positive. Since $h_y = p - 2y$, uniqueness will be guaranteed by the maximum principle provided the solution y remains below $\frac{1}{2}p(t)$. Here, the reduced problem has the two solutions

$$Y_0^\pm(t) = \frac{1}{2}[p(t) \pm \sqrt{p^2(t) + 4}].$$

However, only the solution Y_0^- satisfies the stability condition $h_y(Y_0, t) > 0$. In order to be able to construct a stable initial layer, we therefore restrict the initial values $y(0)$ so that

$$\int\limits_{Y_0^-(0)}^{y} [1 + p(0)r - r^2]dr > 0$$

for all y between $y(0)$ and $Y_0^-(0)$. We similarly restrict $y(1)$ and then we are able to easily construct an asymptotic solution converging to $Y_0^-(t)$ within $(0,1)$.

Exercise

[Compare Howes and O'Malley (1980).]

Consider the two-point problem for

$$\epsilon^2 y'' = h(y, t, \epsilon)$$

where y is an n vector. Show how to formally construct an approximate solution of the form

$$y(t) = Y_0(t) + \xi_0(\tau) + \eta_0(\sigma) + O(\epsilon),$$

where $\xi_0 \to 0$ as $\tau = t/\epsilon \to \infty$ and $\eta_0 \to 0$ as $\sigma = (1-t)/\epsilon \to \infty$, provided

(i) $h(Y_0(t), t, 0) = 0$ and $h_y(Y_0(t), t, 0)$ remains positive-definite throughout $0 \le t \le 1$;

(ii) there exists a scalar function $\phi(\cdot)$ such that $\phi(0) = 0$, $\phi'(0) > 0$, and

$$h'(Y_0(0) + u_0, 0, 0)u_0 \ge \phi(\|u_0\|)\|u_0\|,$$

where the last prime denotes transposition and $\|u_0\| \le \|y(0) - Y_0(0)\|$ in the inner product norm, and where $\int_0^\kappa \phi(r)dr > 0$ for all κ such that $0 < \kappa \le \|y(0) - Y_0(0)\|$; and

(iii) there exists a scalar function $\psi(\cdot)$ such that $\psi(0) = 0$, $\psi'(0) > 0$, and

$$h'(Y_0(1) + v_0, 1, 0)v_0 \ge \psi(\|v_0\|)\|v_0\|$$

for $\|v_0\| \le \|y(1) - Y_0(1)\|$ where $\int_0^\kappa \psi(r)dr > 0$ for $0 < \kappa \le \|y(1) - Y_0(1)\|$.

Example

For the problem

$$\begin{cases} \epsilon^2 y'' = 2y(y^2 - 1), & 0 \le t \le 1, \\ y(0) = -1, & y(1) = 1, \end{cases}$$

the limiting equation has the three solutions $Y_0(t) = 0, -1$, and 1. Indeed, ± 1 are, respectively, upper and lower solutions for our problem, so we are guaranteed that at least one solution exists satisfying $-1 \le y(t) \le 1$. Based on the preceding, we might try to construct a solution $y(t) = 1 + \xi(\tau)$ where $\xi \to 0$ as $\tau = t/\epsilon \to \infty$. Then, however, we would need $d\xi/d\tau = -\xi(\xi+2), \xi(0) = -2$, and this has only the inappropriate constant solution. [We note, however, that such a solution would be appropriate if, instead, $y(0) = -1 + \delta$ for any fixed $\delta > 0$.] Analogously, we cannot construct a solution of the form $y(t) = -1 + \eta(\sigma)$ with a terminal layer. Let us, instead, seek a solution y satisfying

$$y(t) \sim \begin{cases} -1, & 0 \le t < \tilde{t}, \\ 1, & \tilde{t} < t \le 1 \end{cases}$$

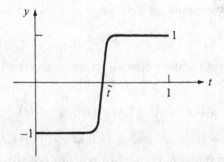

Fig. 3.44. Solution with transition layer.

with a transition layer at $\tilde{t}$ described by $y = z(\kappa)$ for $\kappa = (t - \tilde{t})/\epsilon$ such that $z \to \pm 1$ as $\kappa \to \pm\infty$. Pictorially, in the y–t plane, we would have Figure 3.44. The transition layer solution z would necessarily satisfy $d^2z/d\kappa^2 = 2z(z^2 - 1)$. Integrating from infinity, then,

$$\left(\frac{dz}{d\kappa}\right)^2 = (z^2 - 1)^2$$

so $dz/d\kappa = 1 - z^2$ implies that, up to translation, we must have

$$z(\kappa) = \tanh \kappa.$$

We note that this formally constructed solution seems appropriate for any shock location $\tilde{t}$ within $(0,1)$. This reflects the fact that our differential equation is autonomous. A phase plane argument, however, shows that the solution is unique and described by such a formally constructed solution only for $\tilde{t} = \frac{1}{2}$ [cf. Carrier and Pearson (1968) and O'Malley (1976)]. All nonsymmetric constructed "solutions" for $\tilde{t} \ne \frac{1}{2}$ are (regrettably) spurious.

Let us now seek asymptotic solutions to the more general scalar problem

$$\begin{cases} \epsilon^2 y'' = h(y,t), & 0 \leq t \leq 1, \\ \text{with } y(0) \text{ and } y(1) \text{ prescribed} \end{cases}$$

which involve interior layers. We will suppose the reduced problem

$$h(Y_0, t) = 0$$

has a solution $Y_{L0}(t)$ which is stable [i.e., along which $h_y(Y_{L0}(t), t) > 0$ on $0 \leq t \leq t_L$ and another solution $Y_{R0}(t)$ which is likewise stable on $t_R \leq t \leq 1$ for some $t_R < t_L$]. We will attempt to construct a solution y so that

$$y(t) \to \begin{cases} Y_{L0}(t) & \text{on } 0 < t < \tilde{t}, \\ Y_{R0}(t) & \text{on } \tilde{t} < t < 1 \end{cases}$$

for some $\tilde{t}$ in (t_R, t_L). Because we cannot always expect to have $Y_{L0}(0) = y(0)$ and $Y_{R0}(1) = y(1)$, we will also expect endpoint boundary layers as well. Thus, we impose the usual initial and terminal layer stability assumptions, namely, that

$$\int_{Y_{L0}(0)}^{y} h(r,0)dr > 0 \quad \text{for all } y \text{ between } y(0) \text{ and } Y_{L0}(0)$$

and that

$$\int_{Y_{R0}(1)}^{y} h(r,1)dr > 0 \quad \text{for all } y \text{ between } y(1) \text{ and } Y_{R0}(1).$$

Altogether, then, let us seek an asymptotic solution of the form

$$y(t, \epsilon) = \begin{cases} Y_L(t, \epsilon) + \xi(\tau, \epsilon) & \text{on } 0 \leq t < \tilde{t}, \\ \zeta(\kappa, \epsilon) & \text{on } -\infty < \kappa < \infty, \\ Y_R(t, \epsilon) + \eta(\sigma, \epsilon) & \text{on } \tilde{t} < t \leq 1, \end{cases}$$

where $\tilde{t}$ is to be determined, where $\zeta(\kappa, \epsilon)$ matches $Y_R(\tilde{t} + \epsilon\kappa, \epsilon)$ as $\kappa = (t - \tilde{t})/\epsilon \to \infty$ and $Y_L(\tilde{t} + \epsilon\kappa, \epsilon)$ as $\kappa \to -\infty$, whereas $\xi \to 0$ as $\tau = t/\epsilon \to \infty$ and $\eta \to 0$ as $\sigma = (1-t)/\epsilon \to \infty$. Under our hypotheses, the outer expansions Y_L and Y_R and the boundary layer corrections ξ and η are uniquely obtained termwise, as before. The shock layer solution $\zeta(\kappa, \epsilon)$ must now satisfy

$$\frac{d^2\zeta}{d\kappa^2} = h(\zeta, \tilde{t} + \epsilon\kappa)$$

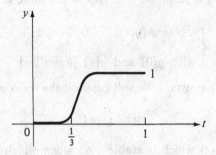

Fig. 3.45. Transition layer solution.

on $(-\infty, \infty)$. Expanding ζ as a power series $\sum_{j=0}^{\infty} \zeta_j(\kappa)\epsilon^j$, the leading term ζ_0 must satisfy the nonlinear equation $d^2\zeta_0/d\kappa^2 = h(\zeta_0, \tilde{t})$. Thus, integration from $-\infty$ implies that

$$\frac{d\zeta_0}{d\kappa} = \pm \sqrt{2 \int_{Y_{L0}(\tilde{t})}^{\zeta_0} h(r, \tilde{t})dr} \quad \text{on} \; -\infty < \kappa < \infty \; \text{with} \; \zeta_0(\infty) = Y_{R0}(\tilde{t}).$$

To reach the desired rest point as $\kappa \to \infty$, we ask that $\tilde{t}$ be an isolated root of the equation

$$H(\tilde{t}) = \int_{Y_{L0}(\tilde{t})}^{Y_{R0}(\tilde{t})} h(r, \tilde{t})dr = 0$$

with, say, $H'(\tilde{t}) \neq 0$ and

$$\int_{Y_{L0}(\tilde{t})}^{y} h(r, \tilde{t})dt > 0$$

for all y values between $Y_{L0}(\tilde{t})$ and $Y_{R0}(\tilde{t})$. We naturally refer to the latter condition as stability of the shock layer. Higher-order terms $\zeta_j(\kappa)$ should follow in a straightforward manner as the solution of successive linear problems. The theory generally follows the work of Fife (1974, 1988).

Example 1

Consider the two-point problem

$$\begin{cases} \epsilon^2 y'' = y(y - a(t))(y - 1), \\ y(0) = 0, y(1) = 1, \end{cases}$$

where $a(t) = \frac{1}{2}t + \frac{1}{3}$. Suppose we wish to find a solution which features a shock layer between the stable solutions $Y_L(t, \epsilon) \equiv 0$ and $Y_R(t, \epsilon) \equiv 1$ at some $\tilde{t}$ in $(0,1)$. The shock layer system $d^2 z/d\kappa^2 = z(z \cdot - \alpha)(z - 1)$ for $\alpha = a(\tilde{t})$ implies upon integration that

$$\frac{dz}{d\kappa} = z\sqrt{\frac{z^2}{2} - \frac{2}{3}(1 + \alpha)z + \alpha}.$$

To have a rest point at $z = 1$ requires us to have $\alpha = \frac{1}{2}$, so the shock layer is described by the simple Riccati equation

$$\frac{dz}{d\kappa} = \frac{1}{\sqrt{2}} z(1 - z).$$

For our linear function a, $\alpha = a(\tilde{t}) = \frac{1}{2}$ implies that the shock must occur at $\tilde{t} = \frac{1}{3}$. Thus, we will have a solution shown in Figure 3.45. Note that this construction would also work for other smooth functions $a(t)$ such that $0 < a(t) < 1$ which cross $a = \frac{1}{2}$. Solutions could then be obtained with a shock layer at any isolated point $\tilde{t}$ where $a(\tilde{t}) = \frac{1}{2}$. [See Kurland and Levi (1988) for a topological study of such problems.]

Example 2

For the problem

$$\begin{cases} \epsilon^2 y'' = y - |t - \frac{1}{2}|, & 0 \le t \le 1, \\ y(0) = y(1) = \frac{1}{2}, \end{cases}$$

it is natural to expect the limiting solution

$$Y_0(t) = |t - \tfrac{1}{2}|$$

throughout $0 \le t \le 1$. Since, however, the shifted Heaviside function $Y_0'(t)$ is discontinuous at $t = \frac{1}{2}$, but satisfies the differential equation elsewhere, we might hope that the solution $y(t)$ for ϵ small will be smoothed out so that it is, at least, continuously differentiable at $t = \frac{1}{2}$. Thus, we will seek an asymptotic solution of the form

$$y(t) = Y_0(t) + \epsilon\xi(\kappa, \epsilon)$$

with $y'(t) = Y_0'(t) + d\xi/d\kappa$ and $y''(t) = (1/\epsilon)(d^2\xi/d\kappa^2)$, etc., where ξ and its derivatives tend to zero as $\kappa = (1/\epsilon)(t - \frac{1}{2}) \to \pm\infty$. We anticipate that the solution and its first derivative will appear as in Figures 3.46 and 3.47. The differential equation implies that

$$\epsilon\frac{d^2\xi}{d\kappa^2} = Y_0 + \epsilon\xi - |t - \tfrac{1}{2}| = \epsilon\xi$$

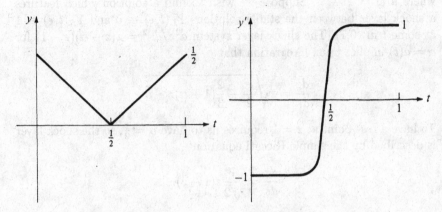

Fig. 3.46. The angular solution.   **Fig. 3.47.** The derivative of the angular solution.

for all κ. Since, however, ξ must decay as $\kappa \to \pm\infty$, we must have

$$\xi(\kappa) = \begin{cases} e^{-\kappa}C_+, & \kappa \geq 0, \\ e^{\kappa}C_-, & \kappa \leq 0 \end{cases}$$

for constants C_+ and C_-. Continuity of y at $t = \frac{1}{2}$ implies that $C_+ = C_-$, whereas continuity of y' then implies that $Y_0'(\frac{1}{2}^-) + C_+ = Y_0'(\frac{1}{2}^+) - C_+$, yielding $C_+ = 1$ and the asymptotic solution

$$y(t) = \left|t - \tfrac{1}{2}\right| + \epsilon e^{-(1/\epsilon)|t-1/2|}.$$

Note that the first derivative is smooth at $t = \frac{1}{2}$, but higher derivatives become unbounded there. In the $\epsilon \to 0$ limit, they indeed act like derivatives of delta-functions.

Appendix

The Historical Development of Singular Perturbations

The history of singular perturbations is quite intertwined with that of a variety of related asymptotic analyses as well as a wealth of applications. Most would agree that the birth of singular perturbations occurred at the Third International Congress of Mathematicians in Heidelberg in 1904, where Ludwig Prandtl spoke "On fluid motion with small friction." His seven page report was contained in the proceedings volume [cf. Prandtl (1905)]. Sydney Goldstein, in reviewing fluid mechanics in the first half of this century [cf. Goldstein (1969)], wrote "In 1928, I asked Prandtl why he kept it so short, and he replied that he had been given ten minutes for his lecture at the Congress and that, being quite young, he thought he could publish only what he had time to say. The paper will certainly prove to be one of the most extraordinary papers of this century, and probably many centuries." Goldstein goes on to point out that the paper went almost, if not completely, unnoticed, perhaps because "it was so very short, and it was published where no one who was likely to appreciate it might be expected to look for it." Prandtl's boundary layer theory (and Blasius' subsequent derivation of the boundary layer equations for low viscosity fluid flow along a flat plate parallel to a stream) received only a single paragraph's coverage in the fifth edition of Lamb's *Hydrodynamics* in 1924, while one section was devoted to the subject in the sixth edition of 1932. More recent surveys of related material are contained in Lagerstrom (1964) and Van Dyke (1964).

The asymptotic integration of linear ordinary differential equations was also being studied early this century; for example, in the thesis of G. D. Birkhoff [cf. Birkhoff (1908)] and in a succession of later work by Noaillon, Tamarkin, Trjitzinsky, Turrittin, Hukahara, Sibuya, and other earlier and later contributors. Quantum theory required the asymptotic solution of Schrödinger's equation via the so-called WKB approximation. The separate analyses of Wentzel, Kramers, and Brillouin, all published in 1926, were actually already contained in those of Liouville and Green from 1837 [cf., the historical surveys of McHugh (1971), Olver (1974), Schlissel (1977), and Wasow (1984)]. Related problems for the fourth-order Orr–Sommerfeld equation (which describes the stability of small disturbances in nonturbulent plane low viscosity fluid flows) have continued to challenge asymptotic analysts [cf. Lin (1955) and Drazin and Reid (1981)]. We simply note the important turning point studies by Gans (1915), Heisenberg (1924), and Langer (1949).

Quite difficult nonlinear oscillation problems began to receive attention in the late 1920s. In particular, van der Pol (1926) studied relaxation oscillations which described triode circuits and he later also analyzed analogous physiological problems [cf. Grasman (1987)]. Generalizing Poincaré's ideas [cf. Minorsky (1947)], Krylov and Bogoliubov and others working in Kiev in the 1930s developed averaging methods to treat problems of fast oscillations [cf. Krylov and Bogoliubov (1947) and Smith (1985)]. A decade later, this fascinating subject and some generalizations were examined by Friedrichs, Levinson, Dorodnicyn, Mitropolsky, and Haag, among others.

Several early mathematical papers concerning singularly perturbed boundary value problems appeared in the 1930s, but they had limited long-term impact. An early such reference is Tschen (1935). Y-W. Chen was a graduate student in mathematics at Göttingen to whom Richard Courant suggested a thesis topic on an ordinary differential equation model featuring boundary layer behavior. Prandtl was professor of aeronautics at Göttingen, and Chen reports that Courant was anxious to encourage a mathematical analysis of boundary layer theory. After Courant fled Nazi Germany, the thesis was completed under Rellich. Chen, however, never did further work on the subject. Motivated by a chemist's question, the prominent Japanese mathematician M. Nagumo studied the initial value problem $\lambda y'' + f(x, y, y', \lambda) = 0$ as $\lambda \to 0$ [cf. Nagumo (1939)]. This followed his very influential 1937 paper on using differential inequalities to obtain solutions to two-point boundary value problems [cf. Howes (1976) concerning its extensive later use for singular perturbations]. Whether Nagumo had learned about Prandtl's theory while he was in Göttingen in the early 1930s is uncertain. Similarly, Rothe (1939) considered the limiting solution of the two-point problem for $y'' + \lambda(y' - y) = \lambda f$ as $\lambda \to \infty$. It followed several earlier asymptotics papers on partial differential equations [cf. Rothe (1933a, 1936)] and two papers on the skin effect in electrical conductors (another boundary layer problem) [cf. Rothe (1933b)], but seems to have also ended Rothe's research on this pregnant topic. Erich Rothe had come to Iowa from Breslau and Berlin, where he might have had some direct exposure to boundary layer theory through his interactions with von Mises.

Wolfgang Wasow was Kurt Friedrichs' first Ph.D. student in the United States. He had left Göttingen and taught high school in Italy before returning to mathematical study in New York through encouragement from Courant [cf. Wasow (1986)]. The term "boundary layer" was borrowed from fluid dynamics and given much greater generality in his substantial thesis [cf. Wasow (1941)]. In addition to Prandtl's work, the thesis was motivated by Friedrichs and Stoker's (1941) study of the buckling of thin elastic plates. {The "edge effect," describing quick changes in stresses and strains along a deformed elastic boundary, had been first studied mathematically for spherical shells by H. Reissner and Blumenthal in 1912 [cf., e.g., Gol'denveizer (1960), Clark (1964), and Ó'Mathúna (1989)], though earlier work by Kirchhoff, Saint-Venant, and Müller-Breslau had certainly anticipated some of it.} The term "singular perturbations" was first used in the title to Friedrichs and Wasow (1946), a paper which followed a productive New York University seminar on nonlinear vibrations. Wasow continued to do groundbreaking work on a variety of asymptotic problems for the next forty years, while Friedrichs was most influential in advancing singular perturbations through his lectures [cf. Friedrichs (1952, 1953, 1955)] and through his encouragement of others' work.

Before long, many further efforts concerning singular perturbations were underway in Cambridge, Moscow, Pasadena, and elsewhere around the globe. Norman Levinson, in particular, generalized the concept of a relaxation oscillation substantially through his definition of discontinuous limiting solutions of nonlinear systems [cf. Levinson (1951)]. By the mid-1950s, he and a number of graduate students and young collaborators at the Massachusetts Institute of Technology (including Aronson, Coddington, Davis, Flatto, Haber, and Levin) made very significant contributions to the asymptotic solution of many kinds of boundary value problems for both ordinary and partial differential equations. It is, indeed, unfortunate that a survey of his work, including its motivations, is unavailable. Tikhonov and his students at Moscow State University, especially Adelaida Vasil'eva, developed exhaustive expansion techniques for wide classes of differential and other equations [cf. Vasil'eva and Volosov (1967), Butuzov et al. (1970), and Vasil'eva and Butuzov (1973)]. That vigorous activity continues to the present time. Other early Soviet workers included Volk, Gradstein, and Mitropolsky. Oleinik's mathematical boundary layer theory and her characterization of generalized solutions through "vanishing viscosity" had great influence in both the East and West [cf. Oleinik (1963, 1968)]. Vishik and Lyusternik's use of boundary layer methods for the asymptotic solution of boundary value problems for linear partial differential (and certain abstract) equations also provided a major step forward [cf. Vishik and Lyusternik (1960, 1961), Trenogin (1970), and Lomov and Eliseev (1988)]. Their methods were, in particular, especially valuable for geometrical optics [cf. Keller (1978)] and for many other important applications. Among the interesting applied singular perturbation problems which migrated westward from the Soviet Union are control theory [cf. Kokotovic et al. (1986)] and semiconductor analysis [cf. Markowich et al. (1990)].

In the late 1940s, a substantial effort developed at the Guggenheim Aeronautical Laboratory of the California Institute of Technology concerning asymptotic solutions of the Navier–Stokes equations. Initial unpublished work included that of Cole and Lagerstrom on viscous compressible fluid flow and Latta's 1951 thesis on a variety of mathematical questions about asymptotic solutions of singular perturbation problems. Cole's (1951) use of Burgers' equation to describe a weak nonsteady shock layer, Kaplun's

(1954) use of appropriate coordinate systems in boundary layer theory, and Lagerstrom and Cole's (1955) asymptotic procedures led to a strong desire to systematically study the asymptotic matching of inner and outer expansions. (Analogous mathematical questions were being simultaneously analyzed in Friedrichs' lectures in New York.) Kaplun and Lagerstrom's work on low Reynolds number flow [cf. Kaplun (1957) and Kaplun and Lagerstrom (1957)] appeared simultaneously with similar work by Proudman and Pearson of Cambridge University. By this time, many others, including Carrier (at Brown), Keller (in New York), and Erdélyi (at Caltech) were solving applied problems and stating theories in the new context of matched expansions and uniformly valid asymptotic approximations. Van Dyke (1970) observed that this was not all new since Maxwell's 1866 study of the viscosity of gases between rotating disks, Kirchoff's 1877 treatment of capacitors, and Laplace's 1805 examination of the meniscus all involved underlying matching arguments. Lagerstrom (1982) argued that matching works through a combination of Kaplun's Extension Theorem and an ansatz concerning the "formal domain of validity," but it is clear that lots of engineers and scientists have successfully solved many applied problems (far beyond fluid mechanics) by using the simplest and most intuitive matching principles. The process was most conveniently presented in Van Dyke's influential 1964 book, *Perturbation Methods in Fluid Mechanics*. Contemporary aerodynamics applications were also discussed in Ashley and Landahl (1965). Wasow's 1965 *Asymptotic Expansions for Ordinary Differential Equations* likewise stimulated considerable mathematical activity concerning singularly perturbed boundary value problems and a much wider spectrum of asymptotic analysis. Various multitime and multiscale methods have been extremely successful in solving problems where local matching is insufficient. Their use in applications (and much more) was greatly influenced by the publication of Cole's 1968 *Perturbation Methods in Applied Mathematics*. Many such specialized techniques also occurred in the various special literatures of mathematical physics, nonlinear oscillations, and wave propagation. Cole and Cook (1986) illustrates anew how pervasive singular perturbation problems remain in modern aerodynamic theory.

Since the mid-1960s, singular perturbations has flourished. The subject is now commonly part of a graduate student's training in applied mathematics and in many fields of engineering. Numerous good textbooks have appeared, and some useful techniques are even found in undergraduate courses and as the basis of common software. The sophistication and power of some of the mathematics used and the breadth of the applications encountered could hardly have been anticipated by those who initiated our study. They would be pleased to acknowledge that important developments are continuing.

References

M. Abramowitz and I. A. Stegen (1965), editors, *Handbook of Mathematical Functions*, Dover, New York.

R. C. Ackerberg and R. E. O'Malley, Jr. (1970), Boundary layer problems exhibiting resonance, *Studies Applied Math.* **49**, 277–295.

R. C. Aiken (ed.) (1985), *Stiff Computation*, Oxford University Press, Oxford.

J. D. Allen and R. E. O'Malley, Jr. (1990), Singularly perturbed boundary value problems viewed in the Liénard plane, *Asymptotic and Computational Analysis* (R. Wong, ed.), M. Dekker, New York, pp. 357–378.

B. D. O. Anderson and P. V. Kokotovic (1987), Optimal control problems over large time intervals, *Automatica* **23**, 355–363.

A. A. Andronov, A. A. Vitt, and S. E. Khaiken (1966), *Theory of Oscillators*, Pergamon Press, London.

T. M. Apostol (1957), *Mathematical Analysis*, Addison-Wesley, Reading, MA.

U. Ascher, R. M. M. Mattheij, and R. D. Russell (1988), *Numerical Solution of Boundary Value Problems for Ordinary Differential Equations*, Prentice-Hall, Englewood Cliffs, NJ.

H. Ashley and M. Landahl (1965), *Aerodynamics of Wings and Bodies*, Addison-Wesley, Reading, MA.

M. Athans and P. L. Falb (1966), *Optimal Control*, McGraw-Hill, New York.

D. J. Bell and D. H. Jacobson (1975), *Singular Optimal Control Problems*, Academic Press, London.

R. Bellman (1970), *Introduction to Matrix Analysis*, 2nd edition, McGraw-Hill, New York.

C. M. Bender and S. A. Orszag (1978), *Advanced Mathematical Methods for Scientists and Engineers*, McGraw-Hill, New York.

A. Bensoussan (1988), *Perturbation Methods in Optimal Control*, Wiley, New York.

S. R. Bernfeld and V. Lakshmikantham (1974), *An Introduction to Nonlinear Boundary Value Problems*, Academic Press, New York.

G. D. Birkhoff (1908), On the asymptotic character of the solutions of certain linear differential equations containing a parameter, *Trans. Amer. Math. Soc.* **9**, 219–231.

J. Bowen, A. Acrivos, and A. Oppenheim (1963), Singular perturbation refinement to quasi-steady state approximation in chemical kinetics," *Chem. Engng. Sci.* **18**, 177–188.

W. E. Boyce and R. C. DiPrima (1986), *Elementary Differential Equations and Boundary Value Problems*, 4th edition, Wiley, New York.

K. E. Brenan, S. L. Campbell, and L. R. Petzold (1989), *Numerical Solution of Initial Value Problems in Differential-Algebraic Equations*, North-Holland, Amsterdam.

G. E. Briggs and J. B. S. Haldane (1925), A note on the kinetics of enzyme action, *Biochem. J.* **19**, 338–339.

R. W. Brockett (1970), *Finite Dimensional Linear Systems*, Wiley, New York.

J. D. Buckmaster and G. S. S. Ludford (1982), *Theory of Laminar Flames*, Cambridge University Press, Cambridge.

V. F. Butuzov, A. B. Vasil'eva, and M. V. Fedoryuk (1970), Asymptotic methods in the theory of ordinary differential equations, *Progress in Mathematics* **8**, Plenum Press, New York, pp. 1–82.

S. L. Campbell (1980), *Singular Systems of Differential Equations*, Pitman, San Francisco.

S. L. Campbell (1982), *Singular Systems of Differential Equations II*, Pitman, San Francisco.

J. Carr (1981), *Applications of Centre Manifold Theory*, Springer-Verlag, New York.

G. F. Carrier (1953), Boundary layer problems in applied mechanics, *Advances in Applied Mechanics III*, Academic Press, New York, pp. 1–19.

G. F. Carrier (1974), "Perturbation methods," *Handbook of Applied Mathematics* (C. E. Pearson, ed.), Van Nostrand-Reinhold, New York, pp. 761–828.

G. Carrier and C. Pearson (1968), *Ordinary Differential Equations*, Blaisdell, Waltham, MA.

G. Carrier and C. Pearson (1976), *Partial Differential Equations*, Academic Press, New York.

J. R. Cash and M. H. Wright (1989), "A deferred correction method for nonlinear two-point boundary value problems: Implementation and numerical evaluation," Computer Science Technical Report 146, AT &T Bell Laboratories, Murray Hill, NJ.

K. W. Chang and F. A. Howes (1984), *Nonlinear Singular Perturbation Phenomena: Theory and Application*, Springer-Verlag, New York.

T. M. Cherry (1950), Uniform asymptotic formulas for functions with transition points, *Trans. Amer. Math. Soc.* **68**, 224–257.

R. C. Y. Chin and G. W. Hedstrom (1991), Domain decomposition: An instrument in asymptotic-numerical methods, *Asymptotic Analysis and the Numerical Solution of Partial Differential Equations* (H. G. Kaper and M. Garbey, eds.), Dekker, New York, pp. 33–54.

R. A. Clark (1964), Asymptotic solutions of elastic shell problems, *Asymptotic Solutions of Differential Equations and their Applications*, (C. H. Wilcox, ed.), Wiley, New York, pp. 185–209.

D. J. Clements and B. D. O. Anderson (1978), *Singular Optimal Control: The Linear-Quadratic Problem*, Springer-Verlag, Berlin.

J. A. Cochran (1962), Problems in Singular Perturbation Theory, doctoral dissertation, Stanford University, Stanford.

J. A. Cochran (1968), On the uniqueness of solutions of linear differential equations, *J. Math. Anal. Appl.* **22**, 418–426.

J. A. Cochran (1972), *Analysis of Linear Integral Equations*, McGraw-Hill, New York.

E. A. Coddington and N. Levinson (1952), A boundary value problem for a nonlinear differential equation with a small parameter, *Proc. Amer. Math. Soc.* **3**, 73–81.

E. A. Coddington and N. Levinson (1955), *Theory of Ordinary Differential Equations*, McGraw-Hill, New York.

J. D. Cole (1951), On a quasilinear parabolic equation occurring in aerodynamics, *Q. Appl. Math.* **9**, 225–236.

J. D. Cole (1968), *Perturbation Methods in Applied Mathematics*, Blaisdell, Waltham, MA.

J. D. Cole and L. P. Cook (1986), *Transonic Aerodynamics*, North-Holland, Amsterdam.

W. A. Coppel (1965), *Stability and Asymptotic Behavior of Differential Equations*, Heath, Boston.

208   References

W. A. Coppel (1975), Linear-quadratic optimal control, *Proc. Royal Soc. Edinburgh* **73A**, 271–289.

W. A. Coppel (1978), *Dichotomies in Stability Theory*, Springer-Verlag, Berlin.

R. W. Cottle (1974), Manifestations of the Schur complement, *Linear Algebra Appl.* **8**, 189–211.

R. Courant and D. Hilbert (1953), *Methods of Mathematical Physics*, Vol. I., Interscience, New York.

J. Cronin (1987), *Mathematical Aspects of Hodgkin-Huxley Neural Theory*, Cambridge University Press, Cambridge.

J. Cronin (1990), Electrically active cells and singular perturbation theory, *Math. Intelligencer* **12**(4), 57–64.

G. Dahlquist (1959), *Stability and Error Bounds in the Numerical Integration of Ordinary Differential Equations*, Trans., Royal Institute of Technology, Stockholm, Nr. 130.

G. Dahlquist (1985), On transformations of graded matrices with applications to stiff ODE's, *Numer. Math.* **47**, 363–385.

G. Dahlquist, L. Edsberg, G. Skollermo, and G. Söderlind (1982), Are the numerical methods and software satisfactory for chemical kinetics?, *Lecture Notes in Math.* **968**, Springer-Verlag, Berlin, pp. 149–164.

G. Dahlquist and G. Söderlind (1982), Some problems related to stiff nonlinear differential systems, *Computing Methods in Applied Sciences and Engineering V* (R. Glowinski and J. L. Lions, eds.), North-Holland, Amsterdam, pp. 57–74.

P. P. N. de Groen (1980), The singularly perturbed turning point problem, *Singular Perturbations and Asymptotics* (R. E. Meyer and S. V. Parter, eds.), Academic Press, New York, pp. 149–172.

K. Dekker and J. G. Verver (1984), *Stability of Runge–Kutta Methods for Stiff Nonlinear Differential Equations*, North-Holland, Amsterdam.

F. Diener (1981), *Méthode du plan d'observabilité*, these, Institute de Recherche Mathématique Avancée, Strasbourg.

F. Diener and M. Diener (1987), *Fleuvres*, C. N. R. S.-Unité Associée No. 212, Paris.

J. Dieudonné (1973), *Infinitesimal Calculus*, Kershaw, London.

J. Dold (1985), Analysis of the early stage of thermal runaway, *Q. J. Mech. Appl. Math.* **38**, 361–387.

F. W. Dorr, S. V. Parter, and L. F. Shampine (1973), Application of the maximum principle to singular perturbation problems, *SIAM Review* **15**, 43–88.

P. G. Drazin and W. H. Reid (1981), *Hydrodynamic Stability*, Cambridge University Press, Cambridge.

W. Eckhaus (1979), *Asymptotic Analysis of Singular Perturbations*, North-Holland, Amsterdam.

A. Erdélyi (1956), *Asymptotic Expansions*, Dover, New York.

N. Fenichel (1979), Geometric singular perturbation theory for ordinary differential equations, *J. Differential Equations* **31**, 53–98.

P. C. Fife (1974), Transition layers in singular perturbation problems, *J. Differential Equations* **15**, 77–105.

P. C. Fife (1988), *Dynamics of Internal Layers and Diffusive Interfaces*, SIAM, Philadelphia.

J. E. Flaherty and R. E. O'Malley, Jr. (1977), The numerical solution of boundary value problems for stiff differential equations, *Math. Computation* **31**, 66–93.

J. E. Flaherty and R. E. O'Malley, Jr. (1980), Analytical and numerical methods for nonlinear singular singularly perturbed initial value problems, *SIAM J. Appl. Math.* **38**, 225–248.

J. E. Flaherty and R. E. O'Malley, Jr. (1984), Numerical methods for stiff systems of two-point boundary value problems, *SIAM J. Sci. Stat. Comput.* **5**, 865–886.

J. E. Flaherty, P. J. Paslow, M. S. Shephard, and J. D. Vasilakis (eds.) (1989), *Adaptive Methods for Partial Differential Equations*, SIAM, Philadelphia.

W. H. Fleming and R. W. Rishel (1975), *Deterministic and Stochastic Optimal Control*, Springer-Verlag, New York.

L. E. Fraenkel (1969), On the method of matched asymptotic expansions, *Proc. Cambridge Phil. Soc.* **65**, 209–284.

K. O. Friedrichs (1952), Special Topics in Fluid Dynamics, lecture notes, New York University, New York.

K. O. Friedrichs (1953), Special Topics in Analysis, lecture notes, New York University, New York.

K. O. Friedrichs (1955), Asymptotic phenomena in mathematical physics, *Bull. Amer. Math. Soc.* **61**, 367–381.

K. O. Friedrichs and J. J. Stoker (1941), The nonlinear boundary value problem of the buckled plate, *Amer. J. Math.* **63**, 839–888.

K. O. Friedrichs and W. Wasow (1946), Singular perturbations of nonlinear oscillations, *Duke Math. J.* **13**, 367–381.

R. Gans (1915), Fortpflanzung des Lichts durch ein inhomogenes Medium, *Ann. Phys.* (4) **47**, 706–736.

T. Geerts (1989), Structure of Linear-Quadratic Control, doctoral dissertation, Technische Universiteit Eindhoven.

A. L. Gol'denveizer (1960), Some mathematical problems in the linear theory of thin elastic shells, *Russian Math. Surveys* **15** (no. 5), 1–73.

S. Goldstein (1969), Fluid mechanics in the first half of this century, *Annual Review Fluid Mechanics* **1**, 1–28.

J. Grasman (1987), *Asymptotic Methods for Relaxation Oscillations and Applications*, Springer-Verlag, New York.

J. Grasman and B. J. Matkowsky (1977), A variational approach to singularly perturbed boundary value problems for ordinary and partial differential equations with turning points, *SIAM J. Appl. Math.* **32**, 588–597.

M. D. Greenberg (1978), *Foundation of Applied Mathematics*, Prentice-Hall, Englewood Cliffs, NJ.

Z.-M. Gu (1987), Analytical Methods for Singularly Perturbed Problems, doctoral dissertation, Rensselaer Polytechnic Institute, Troy, NY.

Z.-M. Gu (1991), A method for solving singularly perturbed systems containing singular manifolds, *Z. Angew. Math. Phys.*

Z.-M. Gu, N. N. Nefedov, and R. E. O'Malley, Jr. (1989), On singular singularly perturbed initial value problems, *SIAM J. Appl. Math.* **49**, 1–25.

S. Haber and N. Levinson (1955), A boundary value problem for a singularly perturbed differential equation, *Proc. Amer. Math. Soc.* **6**, 866–872.

J. K. Hale and K. Sakamoto (1988), Existence and stability of transition layers, *Japan J. Appl. Math.* **5**, 367–405.

G. H. Handelman, J. B. Keller, and R. E. O'Malley, Jr. (1968), Loss of boundary conditions in the asymptotic solution of linear ordinary differential equations, *Comm. Pure Appl. Math.* **21**, 243–261.

W. A. Harris, Jr. (1960), Singular perturbations of two-point boundary problems for systems of ordinary differential equations, *Arch. Rational Mech. Anal.* **5**, 212–225.

W. A. Harris, Jr. (1973), Singularly perturbed boundary value problems revisited, *Lecture Notes in Math.* **312**, Springer-Verlag, Berlin, pp. 54–64.

G. W. Hedstrom and F. A. Howes (1990), Domain decomposition for a boundary value problem with a shock layer, *Domain Decomposition Methods for PDEs* (T. F. Chan, J. Periaux, and O. B. Widland, eds.), SIAM, Philadelphia, pp. 130–140.

F. G. Heinekin, H. M. Tsuchiya, and R. Aris (1967), On the mathematical status of the pseudo-steady state hypothesis of biochemical kinetics, *Math. Biosciences* 1, 95–173.

W. Heisenberg (1924), Über Stabilität und Turbulenz von Flüssigkeitsströmen, *Ann. Phys. Leipzig* (4) 74, 577–627.

P. W. Hemker (1977), *A Numerical Study of Stiff Two-Point Boundary Value Problems*, Mathematical Centre, Amsterdam.

P. Henrici (1962), *Discrete Variable Methods in Ordinary Differential Equations*, Wiley, New York.

F. C. Hoppensteadt (1966), Singular perturbations on the infinite interval, *Trans. Amer. Math. Soc.* 123, 521–535.

F. C. Hoppensteadt (1971), Properties of solutions of ordinary differential equations with a small parameter, *Comm. Pure Appl. Math.* 24, 807–840.

F. A. Howes (1976), "Singular perturbations and differential inequalities," *Memoirs Amer. Math. Soc.* 168.

F. A. Howes (1978), Boundary-interior layer interactions in nonlinear singular perturbation theory, *Memoirs Amer. Math. Soc.* 203.

F. A. Howes and R. E. O'Malley, Jr. (1980), Singular perturbations of semilinear second order systems, *Lecture Notes in Math.* 827, Springer-Verlag, Berlin, pp. 130–150.

A. M. Il'in (1989), *Matching of Asymptotic Expansions of Solutions of Boundary Value Problems*, Nauka, Moscow.

L. K. Jackson (1968), Subfunctions and second-order ordinary differential inequalities, *Advances Math.* 2, Fasc. 3, 307–363.

J. S. Jeffries and D. Smith (1989), A Green function approach for a singularly perturbed vector boundary-value problem, *Adv. Appl. Math.* 10, 1–50.

M. K. Kadalbajoo and Y. N. Reddy (1989), Asymptotic and numerical analysis of singular perturbation problems: A survey, *Appl. Math. Comput.* 30, 223–259.

A. K. Kapila (1983), *Asymptotic Treatment of Chemically Reacting Systems*, Pitman, London.

A. K. Kapila (1989), unpublished lecture notes, Rensselaer Polytechnic Institute, Troy, NY.

S. Kaplun (1954), The role of coordinate systems in boundary-layer theory, *Z. Angew. Math. Phys.* **5**, 111–135.

S. Kaplun (1957), Low Reynolds number flow past a circular cylinder, *J. Math. Mech.* **6**, 595–603.

S. Kaplun (1967), *Fluid Mechanics and Singular Perturbations* (P. A. Lagerstrom, L. N. Howard, and C. S. Liu, eds.), Academic Press, New York.

S. Kaplun and P. A. Lagerstrom (1957), Asymptotic expansions of Navier–Stokes solutions for small Reynolds numbers, *J. Math. Mech.* **6**, 585–593.

D. R. Kassoy (1982), A note on asymptotic methods for jump phenomena, *SIAM J. Appl. Math.* **42**, 926–932.

W. L. Kath (1985), Slowly varying phase planes and boundary layer theory, *Studies Appl. Math.* **72**, 221–239.

W. L. Kath, C. Knessl, and B. J. Matkowsky (1987), A variational approach to nonlinear singularly perturbed boundary value problems, *Studies Appl. Math.* **77**, 61–88.

J. P. Keener (1988), *Principles of Applied Mathematics*, Addison-Wesley, Reading, MA.

J. B. Keller (1978), Rays, waves, and asymptotics, *Bull. Amer. Math. Soc.* **84**, 727–750.

A. Kelley (1967), The stable, center-stable, center, center-unstable, and unstable manifolds, *J. Differential Equations* **3**, 546–570.

W. G. Kelley (1988), Existence and uniqueness of solutions for vector problems containing small parameters, *J. Math. Anal. Appl.* **131**, 295–312.

J. Kevorkian (1990), *Partial Differential Equations*, Wadsworth and Brooks/Cole, Pacific Grove, CA.

J. Kevorkian and J. D. Cole (1981), *Perturbation Methods in Applied Mathematics*, Springer-Verlag, New York.

H. K. Khalil (1981), Asymptotic stability of nonlinear multiparameter singularly perturbed systems, *Automatica* **17**, 797–804.

H. K. Khalil and Y.-N. Hu (1989), Steering control of singularly perturbed systems: a composite control approach, *Automatica* **25**, 65–75.

H. K. Khalil and P. V. Kokotovic (1979), Control of linear systems with multiparameter singular perturbations, *Automatica* **15**, 197–207.

M. Kline (1972), *Mathematical Thought from Ancient to Modern Times*, Oxford University Press, New York.

P. V. Kokotovic (1984), Applications of singular perturbation techniques to control problems, *SIAM Review* **26**, 501–550.

P. V. Kokotovic and H. K. Khalil (eds.) (1986), *Singular Perturbations in Systems and Control*, IEEE Press, New York.

P. V. Kokotovic, H. K. Khalil, and J. O'Reilly (1986), *Singular Perturbation Methods in Control*, Academic Press, London.

N. Kopell (1985), Invariant manifolds and the initialization problem for some atmospheric equations, *Physica D* **14**, 203–215.

N. Kopell (1990), Tracking invariant manifolds in singularly perturbed systems, *SIAM Conference on Dynamical Systems*, Orlando.

V. S. Korolyuk (1990), The boundary layer in the asymptotic analysis of random walks, *Theory Probab. Appl.* **34**, 179–186.

H.-O. Kreiss (1978), Difference methods for stiff ordinary differential equations, *SIAM J. Numer. Analysis* **15**, 21–58.

H.-O. Kreiss (1981), Resonance for singular perturbation problems, *SIAM J. Appl. Math.* **41**, 331–344.

H.-O. Kreiss and J. Lorenz (1989), *Initial-Boundary Value Problems and the Navier–Stokes Equations*, Academic Press, San Diego.

H.-O. Kreiss, N. K. Nichols, and D. L. Brown (1986), Numerical methods for stiff two-point boundary value problems, *SIAM J. Numer. Analysis* **23**, 325–368.

N. M. Krylov and N. N. Bogoliubov (1947), *Introduction to Nonlinear Mechanics*, Princeton University Press, Princeton.

H. L. Kurland (1986), Following homology in singularly perturbed systems, *J. Differential Equations* **62**, 1–72.

H. L. Kurland and M. Levi (1988), Transversal heteroclinic intersections in slowly-varying systems, *Dynamical Systems Approaches to Nonlinear Problems in Systems and Circuits* (F. M. A. Salam and M. Levi, eds.), SIAM, Philadelphia, pp. 29–38.

H. J. Kushner (1990), *Weak Convergence Methods and Singularly Perturbed Stochastic Control and Filtering Problems*, Birkhauser, Boston.

H. Kwakernaak and P. Sivan (1972), *Linear Optimal Control Systems*, Wiley, New York.

P. A. Lagerstrom (1964), Laminar flow theory, *Theory of Laminar Flows*, (F. K. Moore, ed.), Princeton University Press, Princeton, pp. 20–285.

P. A. Lagerstrom (1982), personal correspondence.

P. A. Lagerstrom (1988), *Matched Asymptotic Expansions*, Springer-Verlag, New York.

P. A. Lagerstrom and J. D. Cole (1955), Examples illustrating expansion procedures for the Navier–Stokes equations, *J. Rational Mech. Anal.* **4**, 817–882.

C. G. Lange (1983), On spurious solutions of singular perturbation problems, *Studies Appl. Math.* **68**, 227–257.

R. E. Langer (1949), The asymptotic solution of ordinary linear differential equations of the second-order, with special reference to a turning point, *Trans. Amer. Math. Soc.* **67**, 461–490.

P. D. Lax and C. D. Levermore (1983), The small dispersion limit of the Korteweg-deVries equation I, II, and III, *Comm. Pure Appl. Math.* **36**, 253–290, 571–593, and 809–830.

N. R. Lebovitz and R. Schaar (1975, 1977), Exchange of stabilities in autonomous systems, *Studies Appl. Math.* **54**, 229–259; **56**, 1–50.

J. J. Levin (1957), The asymptotic behavior of the stable initial manifold of a system of nonlinear differential equations, *Trans. Amer. Math. Soc.* **85**, 357–368.

J. J. Levin and N. Levinson (1954), Singular perturbations of nonlinear systems of differential equations and an associated boundary layer equation, *J. Rational Mech. Anal.* **3**, 247–270.

N. Levinson (1951), Perturbations of discontinuous solutions of nonlinear systems of differential equations, *Acta Math.* **82**, 71–106.

A. Liénard (1928), Etude des oscillations entretenues, *Rev. Gen. d'Élect.* **23**, 901–946.

C. C. Lin (1955), *The Theory of Hydrodynamic Stability*, Cambridge University Press, Cambridge.

C. C. Lin and L. A. Segel (1974), *Mathematics Applied to Deterministic Problems in the Natural Sciences*, Macmillan, New York.

X.-B. Lin (1989), Shadowing lemma and singularly perturbed boundary value problems, *SIAM J. Appl. Math.* **49**, 26–54.

J.-L. Lions (1973), *Perturbation Singulières dans les Problèmes aux Limites et en Contrôle Optimal*, Springer-Verlag, Berlin.

S. A. Lomov and A. G. Eliseev (1988), Asymptotic integration of singularly perturbed problems, *Russian Math. Surveys* **43**(3), 1–63.

J. Lorenz (1982a), Stability and monotonicity properties of stiff quasi-linear boundary problems, *Review of Research, Faculty of Science, University of Novi Sad* **12**, 151–175.

J. Lorenz (1982b), Nonlinear boundary value problems with turning points and properties of difference schemes, *Lecture Notes in Math.* **942**, Springer-Verlag, Berlin, pp. 150–169.

J. Lorenz (1984), Analysis of difference schemes for a stationary shock problem, *SIAM J. Numer. Anal.* **21**, 1038–1053.

R. Lutz and M. Goze (1981), *Nonstandard Analysis*, Springer-Verlag, Berlin.

A. D. MacGillivray (1990), Justification of matching with the transition expansion of Van der Pol's equation, *SIAM J. Math. Anal.* **21**, 221–240.

P. A. Markowich (1986), *The Stationary Semiconductor Device Equations*, Springer-Verlag, Vienna.

P. A. Markowich, C. Ringhofer, and C. Schmeiser (1990), *Semiconductor Equations*, Springer-Verlag, Vienna.

B. J. Matkowsky (1975), On boundary layer problems exhibiting resonance, *SIAM Review* **17**, 82–100.

B. J. Matkowsky (1980), Singular perturbations, stochastic differential equations, and applications, *Singular Perturbations and Asymptotics* (R. E. Meyer and S. V. Parter, eds.), Academic Press, New York, pp. 109–147.

J. A. M. McHugh (1971), An historical survey of ordinary differential equations with a large parameter and turning points, *Arch. History Exact Sci.* **7**, 277–324.

L. Michaelis and M. I. Menton (1913), Die Kinetic der Invertinwirkung, *Biochem. Z.* **49**, 333–369.

N. Minorsky (1947), *Introduction to Non-linear Mechanics*, J. W. Edwards, Ann Arbor, MI.

W. L. Miranker (1981), *Numerical Methods for Stiff Equations*, Reidel, Dordrecht.

E. F. Mishchenko (1961), Asymptotic calculation of periodic solutions of systems of differential equations containing small parameters in the derivatives, *Amer. Math. Soc. Translations*, Ser. 2, **18**, 199–230.

E. F. Mishchenko and N. Kh. Rozov (1980), *Differential Equations with Small Parameters and Relaxation Oscillations*, Plenum Press, New York.

P. J. Moylan and J. B. Moore (1971), Generalizations of singular optimal control theory, *Automatica* **7**, 591–598.

J. D. Murray (1977), *Lectures on Nonlinear Differential-Equation Models in Biology*, Clarendon Press, Oxford.

J. D. Murray (1989), *Mathematical Biology*, Springer-Verlag, Berlin.

M. Nagumo (1937), Über die Differentialgleichung $y'' = f(x, y, y')$, *Proc. Phys.- Math. Soc. Japan* **19**, 861–866.

M. Nagumo (1939), Über das Verhulten der Integrale von $\lambda y'' + f(x, y, y', \lambda) = 0$ für $\lambda \to 0$, *Proc. Phys.- Math. Soc. Japan* **21**, 529–534.

A. H. Nayfeh (1973), *Perturbation Methods*, Wiley, New York.

K. Nipp (1988), An algorithmic approach for solving singularly perturbed initial value problems, *Dynamics Reported* **1**, 173–263.

O. A. Oleinik (1963), Discontinuous solutions of nonlinear differential equations, *Amer. Math. Soc. Translations*, Series 2, **26**, 95–172.

O. A. Oleinik (1968), Mathematical problems of boundary layer theory, *Russian Math. Surveys* **23**(3), 1–66.

F. W. J. Olver (1974), *Asymptotics and Special Functions*, Academic Press, New York.

R. E. O'Malley, Jr. (1967), Two-parameter singular perturbation problems for second-order equations, *J. Math. Mech.* **16**, 1143–1164.

R. E. O'Malley, Jr. (1968), Topics in singular perturbations, *Advances in Mathematics* **2**, 365–470.

R. E. O'Malley, Jr. (1970a), On boundary value problems for a singularly perturbed equation with a turning point, *SIAM J. Math. Anal.* **1**, 479–490.

R. E. O'Malley, Jr. (1970b), Singular perturbation of a boundary value problem for a system of differential equations, *J. Differential Equations* **8**, 431–447.

R. E. O'Malley, Jr. (1974), *Introduction to Singular Perturbations*, Academic Press, New York.

R. E. O'Malley, Jr. (1976), Phase-plane solutions to some singular perturbation problems, *J. Math. Anal. Appl.* **54**, 449–466.

R. E. O'Malley, Jr. (1978), Singular perturbations and optimal control, *Lecture Notes in Math.* **680**, Springer-Verlag, Berlin, pp. 170–218.

R. E. O'Malley, Jr. (1980), On multiple solutions of singularly perturbed systems in the conditionally stable case, *Singular Perturbations and Asymptotics* (R. E. Meyer and S. V. Parter, eds.), Academic Press, New York, pp. 87–108.

R. E. O'Malley, Jr. (1983), Shock and transition layers for singularly perturbed second-order vector systems, *SIAM J. Appl. Math.* **43**, 935–943.

R. E. O'Malley, Jr. (1988), On nonlinear singularly perturbed initial value problems, *SIAM Review* **30**, 193–212.

R. E. O'Malley, Jr. (1991), Singular perturbations, asymptotic evaluation of integrals, and computational challenges, *Asymptotic Analysis and Numerical Solution of Partial Differential Equations* (H. G. Kaper and M. Garbey, eds.), M. Dekker, New York, pp. 3–16.

R. E. O'Malley, Jr. and A. Jameson (1975, 1977), Singular perturbations and singular arcs - Parts I, II, *IEEE Trans. Automatic Control* **20**, 218–226; **22**, 328–337.

D. Ó Mathúna (1989), *Mechanics, Boundary Layers, and Function Spaces*, Birkhauser, Boston.

S. Osher (1981), Nonlinear singular perturbation problems and one-sided difference schemes, *SIAM J. Numer. Anal.* **21**, 263–270.

C. E. Pearson (1968), On a differential equation of boundary layer type, *J. Math. and Physics* **47**, 134–154.

D. H. Pollack (1989), Nonlinear, Conditionally Stable, Singularly Perturbed Boundary-Relation Problems, doctoral dissertation, University of Illinois, Urbana-Champaign.

L. S. Pontryagin (1961), Asymptotic behavior of the solutions of systems of differential equations with a small parameter in the higher derivatives, *Amer. Math. Society Translations*, Sec. 2, **18**, 295–319.

P. J. Ponzo and N. Wax (1965), On certain relaxation oscillations: asymptotic solutions, *J. Soc. Ind. Appl. Math.* **13**, 740–766.

L. Prandtl (1905), Über Flüssigkeits - bewegung bei kleiner Reibung, *Verhandlungen, III. Int. Math. Kongresses*, Tuebner, Leipzig, pp. 484–491.

M. H. Protter and H. F. Weinberger (1967), *Maximum Principles in Differential Equations*, Prentice-Hall, Englewood Cliffs, NJ.

R. H. Rand and D. Armbruster (1987), *Perturbation Methods, Bifurcation Theory,* and *Computer Algebra*, Springer-Verlag, New York.

E. L. Reiss (1980), A new asymptotic method for jump phenomena, *SIAM J. Appl. Math.* **39**, 440–455.

G. F. Roach (1982), *Green's Functions*, Cambridge University Press, Cambridge.

E. Rothe (1933a), Über asymptotische Entwicklungen bei Randwertaufgaben der Gleichung $\triangle\triangle u + \lambda^k u = \lambda^k \psi$, *Math. Ann.* **109**, 267–272.

E. Rothe (1933b), Zur Theorie des Skin-effekts, *Z. Physik* **83**, 184–186.

E. Rothe (1936), Über asymptotische Entwicklungen bei gewissen night-linearen Randwertaufgaben, *Comp. Math.* **3**, 310–327.

E. Rothe (1939), Asymptotic solution of a boundary value problem, *Iowa State College J. Sci.* **13**, 369–372.

M. F. Russo (1989), Automatic Generation of Parallel Programs Using Nonlinear Singular Perturbation Theory, doctoral dissertation, Rutgers University, New Brunswick, NJ.

A. Saberi and P. Sannuti (1987), Cheap and singular controls for linear quadratic regulators, *IEEE Trans. Automatic Control* **32**, 208–219.

A. Schlissel (1977), The initial development of the WKB solutions of linear second order ordinary differential equations and their use in the connection problem, *Historia Math.* **4**, 183–204.

C. Schmeiser and R. Weiss (1986), Asymptotic analysis of singular singularly perturbed boundary value problems, *SIAM J. Math. Anal.* **17**, 560–579.

Z. Schuss (1980), *Theory and Applications of Stochastic Differential Equations*, Wiley, New York.

L. A. Segel and M. Slemrod (1989), The quasi-steady state assumption: A case study in perturbation, *SIAM Review* **31**, 446–477.

S. Selberherr (1988), unpublished presentation, BAIL V, Shanghai.

J. G. Simmonds and J. E. Mann, Jr. (1986), *A First Look at Perturbation Theory*, Kreiger, Malabar.

D. R. Smith (1975), The multivariable method in singular perturbation analysis, *SIAM Review* **17**, 221–273.

D. R. Smith (1985), *Singular-Perturbation Theory*, Cambridge University Press, Cambridge.

D. R. Smith (1987), Decoupling and order reduction via the Riccati transformation, *SIAM Review* **29**, 91–113.

J. Smoller (1983), *Shock Waves and Reaction-Diffusion Equations*, Springer-Verlag, New York.

I. Stakgold (1979), *Green's Functions and Boundary Value Problems*, Wiley, New York.

J. J. Stoker (1950), *Nonlinear Vibrations in Mechanical and Electrical Systems*, Wiley-Interscience, New York.

P. Szmolyan (1989), "Transversal heteroclinic and homoclinic orbits in singular pertubation problems," preprint, Institute for Mathematics and its Applications, Minneapolis, MN.

V. A. Trenogin (1970), The development and applications of the asymptotic method of Lyusternik and Vishik, *Russian Math. Surveys* **25**(4), 119–156.

Y. -W. Tschen (1935), Über das Verhalten der Lösungen einer Folge von Differential gleichungen welche im Limes ausarten, *Comp. Math.* **2**, 378–401.

V. I. Utkin (1978), *Sliding Modes and their Application in Variable Structure Systems*, Mir, Moscow.

B. van der Pol (1926), On relaxation oscillations, *Phil. Mag.* **2**, 978–992.

M. Van Dyke (1964), *Perturbation Methods in Fluid Dynamics*, Academic Press, New York. (Annotated edition, Parabolic Press, Stanford, 1975.)

M. Van Dyke (1970), "Nineteenth-century applications of the method of matched asymptotic expansions," unpublished presentation, Conference on Singular Perturbation Techniques, Princeton University, Princeton.

P. M. van Loon (1987), Continuous Decoupling Transformations for Linear Boundary Value Problems, doctoral dissertation, Technische Universiteit Eindhoven.

A. B. Vasil'eva and V. F. Butuzov (1973), *Asymptotic Expansions of Solutions of Singularly Perturbed Equations*, Nauka, Moscow (in Russian).

A. B. Vasil'eva and V. F. Butuzov (1980), Singularly Perturbed Equations in the Critical Case, Report No. 2039, Mathematics Research Center, University of Wisconsin, Madison.

A.B.Vasil'eva and V.F. Butuzov (1990), *Asymptotic Methods in Singular Perturbations Theory*, Vysshaya Shkola, Moscow (in Russian).

A. B. Vasil'eva, A. F. Kardo-Sysoev, and V. G. Stel'makh (1976), Boundary layer in p-n junction theory, *Soviet Phys. Semicond.* **10**, 784–786.

A. B. Vasil'eva and V. G. Stelmakh (1977), Singularly perturbed systems in the theory of transistors, *USSR Comp. Math. Math. Phys.* **17**, 48–58.

A. B. Vasil'eva and V. M. Volosov (1967), The work of Tikhonov and his pupils on ordinary differential equations containing a small parameter, *Russian Math. Surveys* **22**(2), 124–142.

R. E. Vinograd (1952), On a criterion of instability in the sense of Lyapunov of the solutions of a linear system of ordinary differential equations, *Dokl. Akad. Nauk SSSR* **84**, 201–204.

M. I. Vishik and L. A. Lyusternik (1960), The solution of some perturbation problems for matrices and selfadjoint or non-selfadjoint differential equations I, *Russian Math. Surveys* **15**(3), 1–75.

M. I. Vishik and L. A. Lyusternik (1961), Regular degeneration and boundary layer for linear differential equations with a small parameter, *Amer. Math. Soc. Translations*, Ser. 2, **20**, 239–364.

M.J. Ward (1991), Eliminating indeterminacy in singularly perturbed boundary value problems with translation invariant potentials, to be published.

W. Wasow (1941), On Boundary Layer Problems in the Theory of Ordinary Differential Equations, doctoral dissertation, New York University, New York.

W. Wasow (1944), On the asymptotic solution of boundary value problems for ordinary differential equations containing a parameter, *J. Math. and Phys.* **23**, 173–183.

W. Wasow (1965), *Asymptotic Expansions for Ordinary Differential Equations*, Wiley-Interscience, New York.

W. Wasow (1970), The capriciousness of singular perturbations, *Nieuw Arch. Wisk.* **18**, 190–210.

W. Wasow (1984), *Linear Turning Point Theory*, Springer-Verlag, New York.

W. Wasow (1986), *Memories of Seventy Years*, private printing.

M. B. Weinstein and D. R. Smith (1975), Comparison techniques for overdamped systems, *SIAM Review* **17**, 520–540.

E. T. Whittaker and G. N. Watson (1952), *A Course of Modern Analysis*, 4th edition, Cambridge University Press, London.

J. C. Willems, A. Kitapci, and L. M. Silverman (1986), Singular optimal control: A geometric approach, *SIAM J. Control Optim.* **24**, 323–337.

E. Zauderer (1983), *Partial Differential Equations of Applied Mathematics*, Wiley-Interscience, New York.

A. K. Zvonkin and M. A. Shubin (1984), Non-standard analysis and singular perturbations of ordinary differential equations, *Russian Math. Surveys* **39**(2), 69–131.

Index

Applied Mathematical Sciences

cont. from page ii